101+ AI TIPS, TOOLS & PROMPTS

FOR SALES PROS

Sell Smarter and Save Time with Artificial Intelligence

By Julie Holmes

101+ AI Tips, Tools & Prompts for Sales Proessionals

Published by Staten House | Second Edition, 2026

This is a work of nonfiction. While every effort has been made to ensure that the information contained herein is accurate, the author and publisher disclaim any liability for errors or omissions. The advice and strategies contained herein may not be suitable for every situation.

For more information, contact hello@julieholmes.com or visit julieholmes.com.

Staten House

BECOME AN AI-EMPOWERED SALES PRO

The sales landscape is evolving. AI is changing the game, offering practical tools to streamline your workflow, personalize your approach, and close more deals. This guide puts those tools directly into your hands.

What You'll Discover Inside

This comprehensive guide equips you with 101+ actionable tips, useful tools, and effective prompts to:

- **Find better prospects:** Discover AI tools that help you identify and research high-potential leads in minutes instead of hours

- **Write messages that get responses:** Create personalized outreach that sounds like you, not a robot

- **Get your time back:** Hand off scheduling, follow-ups, and data entry to AI so you can focus on actual selling

- **Understand your buyers:** Use AI to spot patterns, anticipate objections, and tailor your approach to each prospect

- **Move deals forward:** Prepare for calls faster, handle objections better, and build proposals that win

How to Use This Resource

Each section is designed for quick action. You'll find prompts you can copy and paste, tool recommendations you can try today, and step-by-step instructions that take minutes to apply. Whether you're new to AI or already experimenting, you'll find strategies that actually work.

Open to any page. Find something useful. Put it to work.

BECOME AN AI-EMPOWERED SALES PROFESSIONAL

Julie Holmes | Smarter HQ
AI Strategist and Tech Founder

Foreword

From Reluctant Seller to AI Convert

I was a reluctant sales professional.

It wasn't something I had ever aspired to do. I was thrust into the role when I worked for a small software startup early in my career. There were parts of the job I excelled at: building relationships with prospects, articulating our value to clients. But there were aspects that overwhelmed, frustrated, and if I'm being honest, terrified me.

Finding just the right words to reconnect with a prospect who had gone silent. Handling tricky negotiations where I felt out of my depth. Keeping all the systems updated with the latest information. Managing communication across email, phone, social media, and in-person meetings. There were never enough hours in the day.

And while I'm no longer a "salesperson" by title, as an entrepreneur and founder, I still sell every day. To clients, to partners, to investors. The challenges remain the same, but the tools have completely changed.

The Research Behind This Guide

When AI burst onto the scene, I wasn't content to just adopt the technology. I needed to understand its true potential for business. I embarked on a journey, interviewing dozens of leaders across industries to discover how organizations were successfully implementing AI, particularly in sales environments.

What I discovered was revealing. The difference between organizations that struggled with AI and those that thrived wasn't budget size or technical expertise. It was their approach.

Through this research, I identified seven fundamental practices that successful organizations consistently followed. These became the UPGRADE Framework:

U **Understand the Potential:** Build a clear, realistic view of what AI can do for your sales process and your results.

P **Position for Impact:** Know where AI creates leverage and where your human selling skills matter most.

G **Grow AI Fluency:** Develop the skills, confidence, and shared language your teams need to use AI well.

R **Reimagine the Business:** Look beyond quick wins to rethink your sales process through an AI-first lens.

A **Activate Tools & Tech:** Select and deploy the right technologies so people can put AI to work immediately.

D **Deliver with Purpose:** Roll out AI changes intentionally so people can adopt and benefit without chaos.

E **Evolve Continuously:** Adapt as capabilities shift, learn from real-world use, and stay ready for what's next.

Today, I share this framework with companies and associations worldwide. The 101+ tips, tools, and prompts in this guide represent what actually works, tested by sales professionals who were once exactly where you are now.

The gap between AI-empowered salespeople and everyone else widens daily. But catching up is easier than you think. You don't need a computer science degree or a massive budget. You just need the right approach.

Your path to AI-empowered sales excellence begins now. Let's go!

Julie

Julie Holmes
AI Strategist | Tech Founder | Global AI Speaker

Table of Contents

JULIE HOLMES
SMARTER HQ

In our AI Readiness Assessment, 99% of leaders say AI will transform their industry... but only 4% believe their organization offers excellent AI training. Meanwhile, 58% report getting no AI training at all.

This skills gap creates a massive competitive advantage for those who master AI strategies now, rather than playing catch-up later.

AI Types and Terms

Learn the Lingo You'll *Actually* Use

Intro to AI

Artificial intelligence isn't just for tech companies (or the tech-savvy) anymore. It's quickly becoming an essential toolkit for sales professionals who want to stay competitive.

At its core, AI helps you work smarter by automating repetitive tasks (like data entry and follow-ups), providing insights you might otherwise miss (like which leads are most likely to convert), and creating personalized content at scale (from emails to proposals).

The best part? AI is the great equalizer in today's market. Small businesses can now use AI tools, agents, and automation to scale their operations, giving them the ability to compete with larger players. Meanwhile, bigger organizations can use AI to deliver the kind of personalization that used to be the hallmark of boutique sales teams.

The key to success isn't just adopting AI tools. It's developing a strategic approach that combines your unique selling strengths with AI's capabilities. That means knowing where AI fits into your workflow, where it doesn't, and how to get the best results without losing your personal touch.

When you use AI effectively, it helps broaden your thinking, take over repetitive work, and speed up tasks while you focus on what humans do best: build relationships, understand complex needs, and provide the genuine connection that closes deals. The sales professionals who figure this out now will have a significant advantage over those still doing everything manually.

Ready to boost your sales performance? Let's dive into the AI concepts and tools that will transform how you sell.

The AI Opportunity Scorecard

The **AI Opportunity Scorecard** is a decision-making tool that helps sales professionals cut through noise and evaluate AI ideas with clarity. Instead of relying on instinct or hype, you score each opportunity against five criteria that reflect real business value. This gives you a consistent way to compare ideas and identify the ones with the strongest potential. The result is a clear, structured path that highlights the opportunities most worth pursuing.

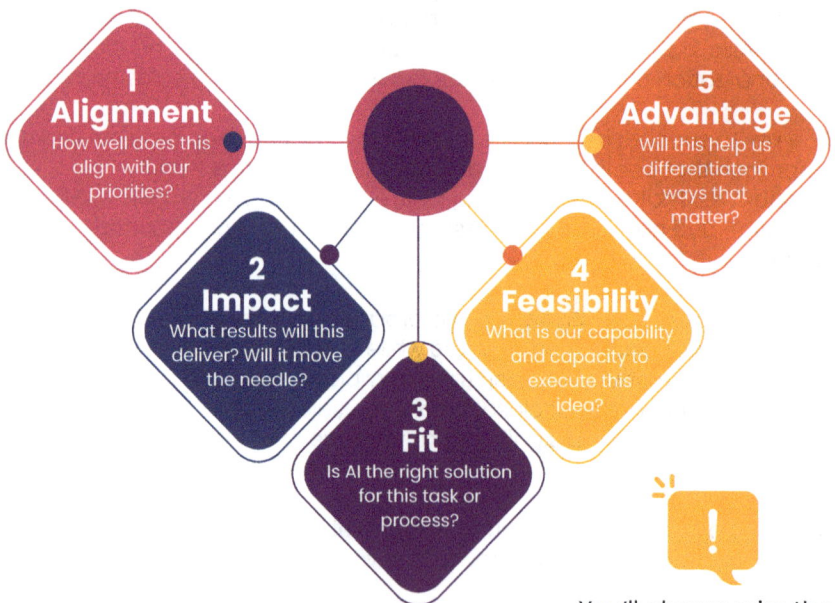

1 Alignment
How well does this align with our priorities?

2 Impact
What results will this deliver? Will it move the needle?

3 Fit
Is AI the right solution for this task or process?

4 Feasibility
What is our capability and capacity to execute this idea?

5 Advantage
Will this help us differentiate in ways that matter?

You'll always miss the right opportunity if you're distracted by the wrong one.

JULIE HOLMES
SMARTER HQ

The AI Opportunity Scorecard

Use the Scorecard

1. **List your AI opportunities**
   Capture every idea you are considering so you can compare them consistently.

2. **Score each idea across all five criteria**
   Use a simple scale such as 1 to 5 for Alignment, Impact, Fit, Feasibility, and Advantage.

3. **Total the scores**
   Add the numbers for each idea to create a clear, side-by-side comparison.

4. **Prioritise the highest scoring opportunities**
   Select the ideas with the strongest overall scores. These are your best candidates for investment and momentum.

Scoring Guide

5 Strong yes. Clearly delivers value, fits perfectly

4 Mostly yes. Some minor gaps but solid overall

3 Mixed. Some potential, but many questions remain

2 Weak. More obstacles than opportunities here

1 Clear no. This does not fit or will not work

Interpreting Your Total

Add the scores across all five criteria. Maximum possible score: 25.

- **20–25: High priority.** Move forward with confidence.
- **15–19: Worth pursuing.** Address the weaker areas before committing.
- **10–14: Proceed with caution.** Resolve significant gaps first.
- **Below 10: Not ready.** Revisit later or redirect your energy elsewhere.

Generative AI

Your Creative Sales Partner

Definition

Generative AI is a type of AI technology that creates new content based on what you ask for. This is the type of AI most often used by sales professionals because it can produce emails, call scripts, proposals, research summaries, and follow-up messages in seconds. Examples of generative AI platforms include ChatGPT, Microsoft Copilot, Claude and Perplexity.

The output you get depends on three variables: the **engine** that processes your request, the **information** you provide, and the **instruction** you give. When these inputs work together, generative AI tools produce clear and relevant results.

Engine
The AI model that processes your request and generates a draft output.

Information
The context and facts you provide so the AI has the right dots to connect.

Instruction
The task, goal, and guidance you give to shape the output you want.

Output
The result the AI produces based on your engine, information, and instruction.

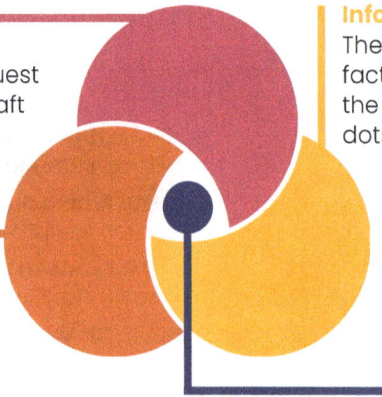

Engine (LLMs/Large Language Models)

What Powers the System

Definition

Large Language Models are the engine behind tools like ChatGPT and Claude. They work by identifying patterns in language and predicting what should come next. The engine does not think like a human. It connects dots based on the inputs you give it. The quality of your output depends on how well the engine can understand your request, which means it needs clear information and instructions to perform at its best.

Example

Ask the engine to 'write a follow-up email to a prospect who mentioned concerns about implementation timeline,' and it will generate a first draft. Add context about your product, their industry, or the tone you want, and the engine produces a more accurate and sales-ready message. The engine supplies speed and processing power. You supply direction and judgment.

Information (AI + Your Company Knowledge)

Giving AI the Right Dots to Connect

Definition

Information is the context, content, and data you give the system so the engine can produce meaningful results. Most AI tools start with pre-trained knowledge from public sources such as websites, books, and research. In enterprise settings, you can also add your own company data in protected environments that maintain confidentiality. This includes documents, plans, metrics, updates, and any details that show how your team or organization operates. When you provide strong information, the AI has accurate building blocks to work with instead of guessing or relying on generic assumptions.

Example

Upload your product brochure, pricing sheet, and a few successful case studies, then ask the AI to 'create a comparison showing how our solution addresses the concerns this prospect raised in our discovery call.' Because the information comes directly from your company, the result is more accurate and tailored to the specific opportunity.

Instruction (Prompting)

Telling AI What You Want

Definition

Instruction is the task you give the system. It includes the details that guide the engine, such as the audience, purpose, tone, and format you want. Sales professionals do not need technical prompting skills; they simply need to communicate clearly. Good instruction creates clarity, and clarity produces better output. We cover this in more depth later in the "Prompt Like a Pro" section.

Example

Instead of saying, 'Write an email,' try: 'Write a 150-word follow-up email to a prospect who requested pricing but hasn't responded in a week. Keep the tone helpful, not pushy, and suggest a quick call to answer questions.' Clear instruction gives the AI a direct path, which leads to a more accurate and relevant result.

Multimodal AI

AI That Understands Images, Audio, and Text

Definition

Multimodal AI can work with more than just written text. It can interpret images, listen to audio, analyze documents, and combine different types of content in a single interaction. For sales professionals, this means the AI can help you analyze a prospect's website, review screenshots of their product or org chart, summarize recorded sales calls, or turn complex inputs into clear insights. Multimodal tools give you fast, integrated perspectives without needing separate apps or manual analysis. This capability is supported in all the major generative AI platforms.

Example

Take a screenshot of a prospect's website and upload it to a multimodal tool. Ask, 'Based on this homepage, what pain points might this company have that our solution addresses?' The AI will examine the visual, interpret the messaging, and offer insights you can use in your outreach. Or try voice prompting to practice your pitch or roleplay objection handling.

AI Agents

Your Digital Sales Assistant

Definition

AI agents are systems that can take initiative and complete multi-step tasks on your behalf. Unlike conversational AI, which responds to a single question at a time, agents can operate more autonomously. They can follow instructions across several steps, interact with other tools or systems, gather information, and complete work in the background. For sales professionals, this means an agent can handle routine workflows such as researching prospects before calls, drafting personalized outreach sequences, updating your CRM after meetings, or preparing proposals without constant supervision.

Example

An AI agent can monitor your pipeline by analyzing deal stages, flagging opportunities that need attention, and drafting follow-up messages for your review. Another agent can prepare proposals by pulling information from previous wins, product specs, and prospect notes, then assembling a tailored first version ready for refinement. These multi-step agents help you maintain speed, accuracy, and consistency across your sales process.

Conversation Analysis

AI Insights From Your Sales Calls

Definition

Conversation analysis tools listen to and analyze your sales calls. They pick up on themes, sentiment, buying signals, objections, and competitive mentions. They help sales professionals understand what resonates with prospects, where deals may be at risk, and how different approaches influence outcomes. All of this happens automatically, without requiring you to review the entire call yourself.

Example

Tools like Fathom, Fireflies, or Otter.ai can record a sales call and highlight moments where your prospect shows interest, raises concerns, or mentions competitors. They can also surface patterns, such as which topics generate the most engagement or which objections come up repeatedly. Use these insights to refine your pitch and improve your close rate.

Predictive Intelligence

Knowing Where to Focus

Definition

Predictive intelligence tools analyze your pipeline and deal data to help you spot patterns and decide where to focus your attention. They identify early warning signs in stalling deals, highlight opportunities most likely to close, and reveal what makes your successful deals work. Sales professionals get a clearer view of where to spend their time, without requiring data science expertise.

Example

Your CRM might flag three deals showing risk patterns similar to past losses, or identify opportunities where buying signals suggest they're ready to move forward. These insights allow you to prioritize the right deals, address concerns early, and stay proactive instead of reactive.

Deep Research & Analysis

Your Competitive Intel Advantage

Definition

AI can turn raw information into actionable sales intelligence in minutes. It reviews industry reports, competitor data, prospect websites, and company news to surface patterns and generate clear takeaways that would normally take hours to compile. Modern models can process large volumes of content at once, making them powerful tools for pre-call research and competitive positioning. Deep research is a separate feature in most AI platforms, so you'll have to actively choose to use it for your analysis.

Example

Before your next big sales call, use the deep research feature and ask, What are the top challenges affecting this prospect's industry this quarter, and how does our solution address them? The tool will search hundreds of sources online, and you can also upload any company reports or competitor updates you have collected. Within minutes, you'll have insights that sharpen your positioning and help you speak their language.

AI Workflow Automation

Connecting Your Sales Tools

Definition

AI workflow automation connects your tools so routine work moves forward without manual effort. These systems coordinate tasks across platforms like Zapier, Make, or built-in automation features inside tools you already use. The goal is simple: less time spent on repetitive steps and more time for actual selling. For sales professionals, automation creates consistency, reduces delays, and ensures that important follow-ups happen even when you're juggling multiple deals.

Example

Imagine a workflow where your meeting AI transcribes sales calls, extracts action items, and sends them straight into your CRM. Your email tool drafts the follow-up message, your calendar schedules the next touchpoint, and your pipeline updates automatically. Everything moves while you stay focused on conversations, relationships, and closing deals instead of administrative tasks.

Team-Facing AI

AI That Works With Your Sales Team

Definition

Team-facing AI is a general description for tools that sales team members interact with directly to answer questions, provide guidance, and reduce day-to-day friction. These can include internal chatbots that handle common inquiries about pricing or product specs, knowledge systems that deliver instant access to sales playbooks and competitive intel, or onboarding assistants that help new reps ramp up quickly. They extend your team's capacity, strengthen consistency, and create a more supported selling experience, even when managers aren't available.

Example

An internal AI assistant can answer rep questions about pricing rules, discount approvals, and product configurations, guide new hires through onboarding tasks, or route more complex requests to the right person. Instead of constant interruptions or unanswered Slack messages, reps get fast, accurate help, and sales leaders gain time back for coaching, pipeline reviews, and strategic conversations.

AI-Enhanced Systems

Smarter Tools Inside Your Existing Workflow

Definition

Many of the tools sales professionals already use now include built-in AI features that automate routine work and surface insights you might otherwise miss. These systems live inside platforms like your CRM, email, chat, or sales engagement tools. They help with data entry, summarization, prioritization, and next-step recommendations, all without requiring a new app or a major rollout. For sales professionals, AI-enhanced systems act like an intelligent assistant inside your existing workflow, reducing administrative load and keeping deals moving.

Example

Tools like Salesforce Einstein, HubSpot AI, or Microsoft Copilot can summarize prospect conversations, draft follow-up emails, highlight deals that need attention, and recommend next steps based on your past wins. Some systems can even draft outreach using your previous messaging as a guide. Everything happens inside the platforms you already use, making it easy to stay on top of your pipeline.

Large language models are trained on hundreds of billions of words of text and code plus images (how else would it know that a cat is a cat?!)

That's equivalent to reading over 83 million books or 700,000 encyclopedias. This massive training allows AI tools to generate context-aware content, solve complex problems, and adapt to your sales needs almost instantly. (Source: Unesco)

Top Tips for Using AI

Strategies to Become an AI-Empowered Sales Pro

JULIE HOLMES
SMARTER HQ

Top Tips for Using AI

AI is becoming an essential capability for sales professionals. It can help you work faster, reduce busywork, and stay on top of your pipeline without adding more hours to your day. When salespeople use AI well, it becomes a strategic partner that boosts prospecting, communication, and execution.

These tips will help you build practical habits that make AI part of the way you sell. They are simple, cross-functional, and designed for busy salespeople who want clarity without complexity. You don't need technical expertise. You only need curiosity, context, and a willingness to experiment.

Think of this section as your quick-start guide to becoming an AI-empowered sales professional. We've included a mix of personal productivity strategies and principles you can apply to prospect smarter, communicate better, and close more deals.

Use these ideas to enhance your thinking, support your sales process, and unlock more time for the work that only humans can do.

Treat AI Like an Intern

TIP # 1

Give Good Guidance to Get Great Results

Imagine AI as a new member of your sales team. It is smart, fast, and eager to support you, but it still needs direction and feedback to produce meaningful work. If you wouldn't let a new hire send a message to your biggest prospect without a quick review, don't let AI do it either. It can't read your mind, and it can't guess the context behind your deals unless you explain it.

Give AI the clarity you'd expect in any team handoff. Tell it who the work is for, why the communication matters, what tone you want, and what outcome you're aiming for. When you articulate the goal, the constraints, and the bigger context, the output becomes sharper and far more aligned with how you actually sell.

This is the foundation of effective collaboration. You set the direction, AI carries the load, and you apply the final polish that adds judgment, nuance, and credibility. Done well, AI becomes a reliable contributor that helps you communicate better, move faster, and stay focused on the work only you can do: building relationships and closing deals.

Talk Your Prospect's Language

TIP #
2

Connect More Effectively

When salespeople ask AI to write an update, they usually get something generic. That's because the request itself is generic. Telling AI to create communication without context is like telling a new rep to go fix the issue without explaining who's involved, what's already been tried, or what outcome you want.

Give AI the same clarity you expect in your organization. Tell it who you're speaking to and what they care about. For example: Draft an email to a VP of Operations who mentioned concerns about implementation timeline during our last call. Reference their specific objection and keep the tone confident but not pushy. Add details about their industry, their role, or what stage of the sales cycle you're in. Writing for a technical buyer? That would probably sound different from a message to a business decision-maker.

These details turn AI from a content generator into a sales partner that understands your environment. The more context you offer about your prospect, their situation, and your goals, the more targeted and meaningful the output becomes.

Write Custom Instructions

TIP #
3

Upfront Effort That Saves Time

Your AI becomes dramatically more effective when it understands who you are and how you sell. That's where custom instructions come in. Your AI profile includes: your tone of voice, communication preferences, role, responsibilities, current priorities, and what good looks like in your world. This information is considered and incorporated for all your future output.

Think of this like onboarding a new sales assistant. If you invest time up front teaching them your expectations, your selling style, and how your organization operates, you get better work with far less explanation later. AI works the same way.

Once saved, your custom instructions reduce repetitive prompting and ensure every output reflects your voice and sales style. Whether you're drafting a prospecting email, preparing for a discovery call, or building a proposal, the AI starts from a much smarter baseline. A one-time setup creates long-term consistency. This small investment now will pay dividends every time you create new content.

Get Your AI Organized

Create Your AI Filing System

Sales professionals generate an enormous amount of AI-supported work: drafts, research, proposals, email sequences, call prep notes, and more. Without structure, your best thinking disappears into a long list of disconnected chats. Treat your AI workspace like an extension of your operating system, not a random stream of messages.

Use projects, folders, or whatever grouping system your AI platform offers to keep related work together. Create a project for each major account, deal stage, campaign, or sales initiative. Add project-level instructions so the AI understands the broader context each time you return. This keeps conversations aligned and dramatically reduces the start from scratch feeling that slows salespeople down.

Rename individual chats clearly so they're easy to find later. A title like Q4 Enterprise Proposal or Acme Corp Discovery Prep makes it simple to continue where you left off and helps the AI stay anchored in the right topic. The more organized your AI work becomes, the more reusable, repeatable, and valuable it will be.

Know When to Upgrade

Match the Tool to the Task

Free AI tools are genuinely useful, and many sales pros get real value from them. But free tiers come with trade-offs: slower responses, shorter memory, limited features, and caps on usage. For occasional tasks or experimentation, free works fine.

Premium versions make sense when AI becomes part of your daily workflow. If you are drafting multiple prospect emails, preparing for back-to-back calls, or building proposals regularly, the upgraded models deliver sharper reasoning, longer context windows, and faster performance that compound into real time savings.

A simple decision rule: Start with free to learn how a tool works. Upgrade when you hit limits that slow you down or when you find yourself working around restrictions instead of just working. If an upgrade saves you even fifteen minutes a day, the ROI is obvious.

Many organizations will cover premium subscriptions for sales teams. Make the business case by tracking the time you save and the quality improvement you see. AI can even help you write that justification.

Review AI Output

Don't Confuse Quantity with Quality

AI can take the heavy lifting out of writing, summarizing, or analyzing, but it still needs your judgment. It works with patterns, not politics, and while it understands language, it doesn't have your lived experience. So, you still need to give everything a thorough review before it leaves your desk.

Read the output like you would review a draft from a capable but new team member. Is the tone right for the moment? Does it match your intent? Is anything missing that your prospect would expect you to address? Are the facts and statistics right (double-check these especially!)?

Be extra careful with anything tied to pricing, competitive positioning, contract terms, or specific client situations. AI can accidentally soften something that needs to be firm or overstate something that needs more care. It can also misinterpret internal terminology or make assumptions that don't fit your deal.

AI speeds up your work, but you still shape the final message. The last review is where your sales instincts show up, and where you make sure the output reflects your standards, your voice, and the relationship you're building.

Use More Than Text

TIP #

7

Go Multi-Modal

Sales professionals handle information in every format imaginable: screenshots, charts, documents, slides, prospect websites, proposal markups, and the occasional urgent message sent from a parking lot before a meeting. Modern AI tools can work with all of it, and using these features saves you hours.

Instead of typing long explanations or trying to describe a messy document, upload the image, document, or file directly. The AI can analyze structures, extract themes, summarize key points, or give you options for how to communicate something to your prospect. It's far easier for the system to understand your world when you show it the actual materials you work with.

Want to level up even more? Use dictation to describe something on the fly. Or voice mode (with or without video) to describe something or role-play a conversation. These options can save you time and give you a great way to share imperfect thoughts or messy ideas.

All of this can give AI a clearer picture of your environment, which means you spend less time explaining and more time selling. When you use every input available, AI becomes far better at supporting your day-to-day deals.

Analyze Your Meetings

Learn More from Every Conversation

Sales professionals spend a huge portion of their time in meetings, but most of the insight from those conversations evaporates the moment the call ends. AI can help you capture and analyze what actually happened so you can sell with more clarity and intention.

After a meeting, feed the transcript or recording into your AI tool and ask for patterns. What themes came up repeatedly? Where was there tension? What decisions were made? What concerns surfaced that no one followed up on? You can even ask, Based on this conversation, what should I pay attention to as a next step?

This kind of analysis helps you see what you might miss in the moment. Maybe a prospect raised a concern three times. Maybe they kept circling back to pricing. Maybe a stakeholder who usually contributes stayed quiet. These signals matter.

Using AI to analyze your meetings gives you a clearer understanding of your prospect's priorities, objections, and pressure points. It also helps you prepare better for follow-ups, coach yourself more effectively, and keep deals moving without guesswork.

Automate the Admin

Sell More by Typing Less

Administrative work drains selling capacity. Meeting summaries, follow-up notes, CRM updates, weekly forecasts, and status reports can quietly consume hours of your week. These tasks matter, but they rarely require your judgment or strategic thinking. This is where AI shines.

Use AI to draft your summaries, organize action items, prepare follow-up emails, and create first-draft updates. Offloading this repetitive writing gives you time back for the work that actually moves your pipeline forward. The question to ask is simple: Does this task require my selling skills, or could AI create the first version?

The more you delegate the repetitive writing, the more energy you have for prospecting, discovery, and closing. Many sales professionals underestimate the compounding effect of this shift. That difference doesn't just save time. It creates space for clearer thinking, stronger communication, and much better use of your selling attention. AI cannot build relationships or close deals, but it can free you to do more of both.

Look for AI Integrations

TIP #
10

Get More from the Tools You Already Use

You don't always need new software to get value from AI. Many of your existing tools already include AI features that can streamline workflows and improve productivity. These built-in enhancements are often the easiest place to start because there is no learning curve and no new system to adopt.

Keep an eye on updates from your CRM, email platform, sales engagement tools, and communication apps. Most of them are adding AI capabilities that summarize conversations, surface action items, draft messages, organize tasks, and highlight what needs your attention.

Because these upgrades sit inside tools your team already uses, adoption is immediate. When Salesforce added AI summaries and HubSpot rolled out AI-assisted email drafts, sales teams found that their daily noise dropped dramatically. They didn't need training or a new process. Their existing tool simply became more helpful.

These small integrations create big wins. They save time, reduce manual effort, and build confidence in AI without introducing complexity.

Start with High-Value Low-Risk

TIP #
11

Choose Your First AI Wins Wisely

When you are getting started with AI, don't begin with your most sensitive or high-stakes work. Start with tasks that have meaningful value but low risk if they are not perfect. Meeting preparation, prospect research, brainstorming, and first-draft communication are ideal early wins. They build your skills, strengthen your confidence, and help you understand AI's strengths without pressure.

These early projects give you room to experiment without pressure. Once you feel comfortable guiding the AI and refining its output, you can move to more complex work like strategy documents, proposals, or important client messaging.

Sales professionals across industries follow this progression naturally. Consider it your practice field. Build your capability where the stakes are low so you're ready when the work is high-impact. Smart salespeople start small, learn fast, and scale with confidence.

Create Team Guidelines

TIP #
12

Set Boundaries Without Killing Innovation

Your sales team needs clarity about how to use AI responsibly, but the last thing you want is a fear-based policy that shuts down experimentation. Sales leaders set the tone here. If your guidance feels like a legal disclaimer, people will avoid using AI altogether. If it feels clear and practical, they'll use AI confidently and appropriately.

Keep your guidelines simple and focused on the essentials: data privacy, when to use AI versus a human, and the review process for AI-generated work. A helpful starting point is: Use AI for drafts, research, and call prep. Don't upload sensitive client information or confidential deal terms into public tools. Review AI output before it's shared outside the team.

Your goal is to give people permission to explore while protecting your organization. A short, well-written guide helps your team understand what's safe, what's encouraged, and what requires oversight. It also reduces the invisible hesitation many people feel when trying new technology. The easier you make it to use AI safely, the faster your team builds collective skill and confidence.

JULIE HOLMES
SMARTER HQ

Share Your Journey

TIP #
13

Vulnerability Builds Trust

You don't need to pretend you're an AI expert. In fact, acting like you've mastered everything can unintentionally shut down learning in your team. When sales leaders openly share what they're trying, what worked, and what fell flat, it creates psychological safety. It tells your team that experimentation isn't just allowed, but expected.

Talk openly about your own AI experiences. Mention when you used AI to draft a prospecting sequence, synthesize competitor research, or prepare for a big pitch. Share the moments when the output surprised you and the times when it completely missed the mark. This honesty normalizes the learning curve.

When sales leaders model curiosity instead of perfection, teams feel more comfortable taking risks. They surface ideas faster, admit when they get stuck, and collaborate more effectively. This openness accelerates skill-building across the entire sales organization.

Your goal is not to be the AI expert in the room. Your goal is to show that you're learning alongside your team and leading the way by example. A transparent learning journey builds trust, momentum, and a culture where innovation feels safe.

Use AI for Strategy

Go Beyond Task Completion

Most salespeople start with AI for tactical tasks: drafting emails, summarizing calls, or creating proposals. Those are good entry points, but they barely scratch the surface of what AI can help you do. The real value emerges when you use AI as a thinking partner, not just a task assistant.

Ask AI to challenge your deal strategy, pressure-test your assumptions, or highlight blind spots. Try prompts like What risks am I not considering?, Which stakeholders might resist this purchase and why?, or Generate three alternative approaches that could achieve the same outcome.

This shifts AI from a drafting tool to a strategic ally. It helps you surface issues earlier, explore a wider range of possibilities, and strengthen your decision-making before you walk into the room.

AI won't make strategic decisions for you, but it can help you think more deeply and prepare more thoroughly. Use it to explore scenarios, play out consequences, and sharpen your plans. It's better to find weaknesses during preparation than during a critical meeting.

Build AI Champions

TIP #
15

Create Your AI Advocacy Network

Driving AI adoption shouldn't be a solo effort. You need advocates across the sales organization who model good use, share wins, and help others build confidence. Every team has natural early adopters who get curious about new tools and enjoy helping colleagues. These are your AI champions.

Identify the people who are already experimenting, asking smart questions, and finding ways to make their work easier with AI. Give them space to share what they are learning in team meetings, internal channels, or short demos. When peers show how AI saves time or improves work quality, it normalizes experimentation far faster than top-down directives ever could.

Your role is to create the conditions for this peer-led momentum. Recognize their efforts, amplify their ideas, and encourage teams to share small victories. Over time, you build a distributed network of advocates who help others learn, reduce hesitation, and make AI part of the culture rather than a management mandate. Champions accelerate adoption. And they turn AI from a sales initiative into an organizational capability.

Ask AI to Challenge You

Test Your Thinking for Better Decisions

Sales professionals make decisions under pressure, often with limited time and incomplete information. It is easy to default to familiar patterns or overlook assumptions you have been carrying for years. AI can help you pressure-test your thinking by challenging your assumptions and offering alternative viewpoints.

Before making a significant decision, describe your reasoning to AI and ask direct questions such as, What assumptions might be wrong here? or What alternative explanations could fit this data? This pushes you to consider angles you may have dismissed and brings hidden risks or blind spots to the surface.

You are not outsourcing the decision. You are strengthening it. AI gives you a rehearsal space to explore weaknesses quietly before they become public problems. Sometimes it will surface a valid concern you had not considered. Other times it will confirm that your thinking is sound. Both outcomes help you decide with more confidence and clarity.

Asking AI to challenge your logic doesn't replace your judgment. It enhances it.

Create Your AI Routine

Make AI Part of How You Work

AI delivers the most value when it becomes part of your day-to-day sales rhythm, not an occasional tool you remember to try. Build small, consistent habits that integrate AI into the way you prospect, communicate, and manage your pipeline.

Start your morning by asking AI to review your calendar and surface what matters most based on deal stage, urgency, and preparation needed. Before each meeting, take two minutes to have AI summarize past notes or identify likely objections. At the end of the day, ask AI to recap what you accomplished and outline the top three priorities for tomorrow.

These micro-routines compound into real benefits: stronger preparation, faster follow-up, clearer communication, and significantly less administrative drag. You spend less time catching up and more time selling.

Think of AI not as another step or task, but as a quiet partner sitting beside you throughout the day. Small, steady use builds familiarity and confidence, and over time, it changes how you think, prepare, and operate.

Build an AI Decision Tree

Know When to Use AI and When to Be Human

Effective sales professionals develop a clear set of principles for when to lean on AI and when human judgment is essential. Without a simple framework, you risk overusing AI where it does not belong or underusing it where it could save you hours.

Use AI for tasks that involve research, first drafts, summarization, idea generation, data analysis, pattern spotting, and scenario exploration. These are areas where speed and breadth can add more to your selling effectiveness.

Use humans for anything involving relationships, sensitive negotiations, complex client situations, or deep account context. These require empathy, nuance, and trust.

Most work falls somewhere in between. Over time, you'll build the judgment to decide which approach fits the moment. A good rule of thumb is: if the task strengthens relationships or requires lived experience, humans lead. If it is repetitive, analytical, or heavily informational, AI leads, but humans review.

Stay Current with AI

TIP # 19

Embrace the Learning Curve

AI is evolving rapidly. New features, new models, and new capabilities emerge every month, not every year. Sales professionals who stay curious and keep learning will outpace those who wait for things to stabilize. There is no stable. This is the new normal.

Set aside a small, consistent block of time each week to explore something new. Try a new feature. Test a new prompt. Watch a quick demo. Read a short article. The goal is not mastery. The goal is staying aware of what is possible so you can stay ahead of your competition.

Encourage your fellow salespeople to do the same. Share insights during team meetings. Celebrate experimentation and keep the conversation alive. When sales teams stay current, the whole organization feels permission to learn and adapt.

You don't need to become an AI expert. You simply need to stay engaged. Curiosity is now one of your most valuable selling skills.

Respect Privacy and Ethics

Use AI Responsibly

As a sales professional, you set the standard for how AI is used across your deals and client relationships. That starts with protecting sensitive information. Be thoughtful about what you upload into AI tools, especially when it involves prospect data, pricing details, contract terms, or anything confidential. Public AI tools are not always the right place for that data.

Check your organization's AI policy before sharing internal documents. If you're unsure, swap sensitive details for placeholders and fill in the real information later. It is a simple habit that prevents unnecessary risk.

Responsible AI use builds trust. Your team needs to know you take privacy seriously. Your organization needs confidence that you understand the boundaries. And your customers expect that their information is handled with care.

Ethics isn't a footnote in AI adoption; it's the foundation. The way you use AI teaches others how to use it. Model responsibility, and your teams will follow.

75% of knowledge workers are already using AI at work, and 78% are bringing their own AI tools to do it.

People aren't waiting for permission. Shadow AI use is now one of the biggest risks for organizations, not because AI is dangerous, but because using it without clear guidance is.
(Source: Microsoft)

Prompt Like a Pro

Master the Art & Science of AI Communication

BE PREPARED

Understanding the Prompt Framework

TIP # 21

ChatGPT and similar tools are transforming AI interactions, turning your ideas into detailed responses.

The key to unlocking its full potential? The prompt (instructions) you provide. Introducing the **PREPARED framework**, your ultimate guide to crafting effective prompts.

The **PREPARED framework** outlines essential elements to consider when creating prompts, ensuring you get the most accurate and relevant responses. This framework is your key to mastering the crucial first 20% of the AI collaboration process, where your input shapes everything that follows.

While not every element is necessary for every prompt, incorporating them can significantly enhance the quality of the results. Remember, don't skimp on the details. The more information you provide to any generative AI tool, the better the response you'll receive.

Remember: Prompting is only part of collaborating effectively with AI. Keep reading to understand the full Collaboration Equation (20-60-20).

PREPARED Framework: Cheat Sheet

P **Persona:** Define who AI should act as and the viewpoint it should take so responses match the right role, expertise, and tone.

R **Request:** State clearly what you need produced - email, analysis, script, plan - so AI knows the exact task.

E **Explain:** Give the context, background, and why this matters. The "story behind the task" drives better outputs.

P **Process:** Lay out the steps or structure you want AI to follow so the response is organized the way you need it.

A **Aim:** Describe the intended outcome and what success looks like when you use this output - to persuade, encourage, inform, etc.

R **Restrictions:** Set boundaries or must-haves, word limits, tone, topics to avoid, compliance rules, or details that must be included.

E **Examples:** Add sample phrasing, formats, or styles you want AI to imitate so the result feels familiar and on-brand.

D **Discuss:** Review or refine the work with AI after you get your first draft by asking questions, surfacing missing information, requesting alternatives, or challenging assumptions.

Persona

Dress Your AI for Success

P

Definition

AI contains 1000s of perspectives - which one do you want? Think of setting a persona for AI as giving it a role to play, just like how you might adjust your tone when presenting to the board versus coaching a direct report. By telling AI who you want it to be, you get responses tailored to that expertise. The right persona transforms generic advice into insights that actually fit your leadership context.

Example

Instead of starting with "Help me write an email", try: "Act as an experienced sales director who has closed over $5M in SaaS deals. Draft an email to our champion summarizing the business case they can take to their CFO."

For discovery calls: "Act as a consultative sales expert who asks insightful questions." For competitive situations: "Act as a market research analyst who specializes in competitive intelligence." The persona you choose sets the foundation for everything that follows.

Request

Crafting a Clear Call to Action

R

Definition

Your request tells AI exactly what you want it to create or do. Being precise here is like giving clear direction to a talented but inexperienced team member to eliminate confusion and get you better results faster. Vague requests get vague responses. Specific requests get outputs you can actually use.

Example

Instead of "Help me with a follow-up email", try: "Draft a follow-up email to a prospect who missed our scheduled demo call yesterday but previously expressed interest in our automation features."

For discovery call prep: Instead of "Help me prepare for a call," try: "Create 5 discovery questions for a first call with a VP of Operations at a manufacturing company that will help me understand their inventory management challenges."

For proposals: "Develop a one-page executive summary for a prospect who is comparing us against two competitors and is most concerned about implementation timeline."

Explain

Setting the Scene with Context

E

Definition

Context is king. The more background you provide about your organization, team dynamics, and current situation, the more tailored and useful AI's response will be. Skip this step, and you'll get generic content that usually requires heavy editing or a lot of back and forth.

Example

"We sell a marketing automation platform to mid-market e-commerce companies. Our prospect is the VP of Marketing at a 200-person retailer currently using basic email marketing but struggling with personalization at scale. They've had two demos with us, expressed interest in our dynamic content features, but raised concerns about implementation time and internal resources. Their contract with their current vendor expires in 60 days."

Or: "I lead a 40-person sales team selling HR software. We're entering Q4 and need to close 15 more enterprise deals to hit our annual target. Half the team is ahead of quota, the other half is struggling with longer sales cycles than expected. Leadership wants AI integrated into our sales process, but we can't agree on where to start."

Process

Charting the Course

P

Definition

When you need AI to follow a specific workflow or include specific output headings/sections, outline the process step by step. Think of this like writing a management report where you start with a template and fill in the blanks. This approach can be valuable for complex leadership tasks like strategic planning, change management, or cross-functional initiatives where sequence matters. In short, you can use this to create an SOP for your AI collaborator.

Example

"To create this proposal for our enterprise prospect:
- Start with their three stated priorities and how our solution addresses each one
- Provide concrete metrics showing similar customer results (I'll give you the data), but you should ask if anything appears to be missing or needs more context
- Address the two main concerns they raised: implementation timeline and integration with their existing tech stack
- End with our recommendation for a phased rollout, including pricing options and risk mitigation."

Aim

Clarifying the End Goal

A

Definition

Aim is one of the most valuable elements of a prompt and often the one that's missing. Sharing this in a prompt can drastically guide tone and approach for the AI. What's the actual outcome you're trying to achieve? Initial ideas or polished final version? Persuade executives, inform your team, or facilitate discussion? Being explicit about your aim helps AI pick the best words, depth, and structure. Skip this and you'll get well-written content that's totally wrong for your purpose.

Example

"My goal is to get this prospect to agree to a pilot program. This isn't about closing the full deal yet, just getting them comfortable enough to test our solution with one team before expanding company-wide."

Or: "I really want you to help me think through this deal strategy, not make a recommendation. I need to see the pros and cons of offering a discount versus adding more value to justify our pricing. Surface considerations I might be missing and challenge my thinking."

Restrictions

Setting Requirements and Boundaries

R

Definition

Think of this element as articulating anything you definitely DO or DO NOT want the AI to do. This includes specifying tone, length, format constraints, concepts to include and what to avoid. This is especially important for maintaining your leadership voice, adhering to company culture, and ensuring compliance with policies. Think of restrictions as guardrails that keep AI's creativity focused on what actually works in your organization.

Example

"Keep the proposal under 5 pages since this prospect prefers short, focused documents. Avoid AI buzzwords like 'transformational' or 'revolutionary.'"

Or: "Provide source links for any statistics or claims. Factual accuracy and message integrity are key, so don't make up stories."

For regulated industries: "No claims about performance improvements or cost savings without customer data backing it up. We're in financial services, so everything must be defensible. Don't mention specific competitors as the prospect hasn't brought them up."

Examples

Showing is Better Than Telling

E

Definition

Providing samples of what you expect gives AI a clear model to follow. Whether it's your communication style, a successful memo, or a board report structure you liked, examples crystallize expectations better than descriptions alone. This is especially powerful for matching your company culture or personal leadership voice.

Example

"Here's how I typically communicate with prospects at this stage: 'Thanks for making time yesterday. I know implementation timeline is top of mind for your team, so I wanted to share how a similar company got up and running in 6 weeks. Happy to walk through the details if helpful.' Use a similar direct, helpful approach for this follow-up email."

Or: "Our top rep starts discovery calls by asking about business impact before diving into features. Match that approach for these discovery questions."

JULIE HOLMES
SMARTER HQ

Discuss

The Art of Iteration

D

Definition

Getting great results requires back-and-forth refinement. Give feedback, ask for revisions, or request explanations of AI's choices. This collaborative approach produces the best results. Remember: AI is your intern, and good managers offer feedback and coach to get to excellence.

Example

After receiving a draft: "This is 80% there, but too formal for this prospect. Make it conversational like how I might talk on a discovery call. Also, you focused on features but didn't tie them to their specific pain points. Add a 'What This Means for You' section with 2-3 specific outcomes."

Or: "Explain your reasoning. Why lead with ROI instead of implementation speed? I need to understand the logic before sending to this prospect."

For competitive situations: "This battlecard feels too defensive. Translate it into confident positioning. Instead of 'addressing their concerns,' help me reframe around our strengths. Use that type of approach throughout."

PREPARED in Action
Draft an Email

P **Persona:** You're an experienced sales pro who specializes in re-engaging stalled deals and moving prospects to next steps.

R **Request:** Draft an email to a prospect (VP of Operations) who went dark after receiving our proposal two weeks ago.

E **Explain:** We had three strong discovery calls and they said our solution was their top choice. Then silence after we sent pricing. Their fiscal year ends in 6 weeks.

P **Process:** Acknowledge the gap in communication without being pushy. Offer a new angle or piece of value. Invite them to a brief call to address any concerns.

A **Aim:** The goal of the email is to get a response and re-open the conversation, not to close the deal in this message.

R **Restrictions:** Keep it under 150 words. No guilt-tripping language. Avoid phrases like 'just checking in' or 'circling back.' Keep tone confident but respectful.

E **Examples:** Here's a sample for reference, "Our goal is to help you hit your Q4 targets, and I have a few ideas on how we could accelerate implementation if timeline is a concern."

D **Discuss:** After generating the initial email draft, explain which approach you took and why. Ask for help identifying phrases that might feel too salesy, or suggest an alternative angle if this one doesn't get a response.

PREPARED in Action
Team Feedback Support

P — **Persona:** You're an executive coach helping a sales manager prepare fair, valuable performance reviews for a 12-person team.

R — **Request:** Draft talking points I can use during each review to reinforce expectations, highlight strengths, and address development needs.

E — **Explain:** We're launching AI chatbots for Tier 1 customer service tickets. Half our 25-person team is excited, half worried about job security.

P — **Process:** Give me 1) a quick summary of goals, wins, and gaps and 2) two strengths and one development priority with examples and 3) a list of possible questions that might come up.

A — **Aim:** My goal is to lead consistent, confident reviews that feel personalized rather than generic with less prep time.

R — **Restrictions:** Avoid corporate clichés. Keep each set of talking points under 150 words. No promises about promotions or compensation. Keep tone direct and supportive.

E — **Examples:** Here's how I might say something: "We just finished Q3, and I need to deliver reviews that address both quota performance and skill development. Half exceeded target, the other half struggled with pipeline generation."

D — **Discuss:** After reading the draft, you might ask: "Does this reflect my authentic voice?" or "What might my direct report be thinking after seeing feedback like this?" "Adjust for this sensitivity, cultural context, or context from past conversation."

Collaborate with AI

Partner Like a Pro

Now that you've mastered the PREPARED framework for crafting effective prompts, let's see how this fits into the bigger picture of working with AI. The 20-60-20 framework shows you how great prompting leads to a complete, efficient AI workflow that saves you time while maintaining quality and your personal touch.

The biggest mistake sales pros and leaders make with AI? Either expecting it to do everything perfectly or trying to do everything themselves. This formula will help ensure you strike just the right human-AI balance. It all starts and ends with YOU.

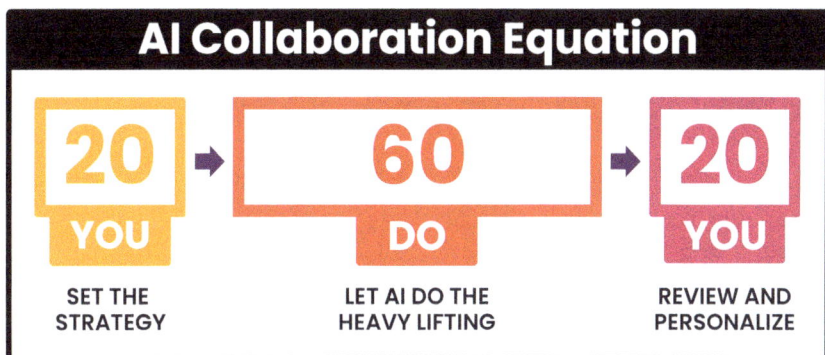

AI Collaboration Equation

20	60	20
YOU	**DO**	**YOU**
SET THE STRATEGY	LET AI DO THE HEAVY LIFTING	REVIEW AND PERSONALIZE

Remember: The quality of your information and instructions (using PREPARED) determines the quality of AI's output. This is all part of your crucial first 20% in the AI collaboration process.

The 20-60-20 Collaboration Equation

First 20% → YOU → Set the Strategy

Give AI clear direction by crafting strong prompts using the PREPARED framework, choosing the right tool, and providing necessary context.

Middle 60% → AI → Does the Heavy Lifting

Let AI draft emails, create presentations, analyze data, generate strategic frameworks, and handle time-consuming tasks.

Final 20% → YOU → Review and Add Your Human Touch

Add your leadership and/or sales expertise, check for accuracy, adjust tone, and include personal insights that only you would know about your team and stakeholders.

Why this equation matters:

Ever spotted content that screams "AI-generated"? The generic phrasing, awkward transitions, and hollow tone are immediate giveaways.

This happens when people skip the critical first 20% (poor prompting) or neglect the final 20% (personalization). They paste vague requests into their AI tool, take whatever comes out, and wonder why it sounds robotic.

The difference between AI that enhances your voice versus replacing it comes down to mastering this framework. When you invest in crafting specific prompts with clear context and then properly personalize the output, the results become indistinguishable from your best work - only created in a fraction of the time.

20-60-20 in Action
Real-World Examples

Prospecting

20%
You identify target accounts and key details
↓
60%
AI researches and drafts personalized outreach
↓
20%
You refine tone and add insights only you would know

Proposal Writing

20%
You outline requirements and win themes
↓
60%
AI generates a structured proposal draft
↓
20%
You customize pricing and add client-specific details

Call Prep

20%
You define what you need to know
↓
60%
AI creates a briefing wiht talking points and potential objections
↓
20%
You prioritize based on your knowledge of the prospect

Pro Tip: As you build skill and confidence, you can shift toward the 10-80-10 model, where AI handles the majority of the workload. You'll know you're ready when reviewing AI output feels like polishing rather than rewriting, allowing you to spend your final 10% adding the human insights that close deals.

Common Prompting Mistakes
Avoiding Prompt Pitfalls

1. **Being too vague:** "Write me an email" vs. "Write a follow-up email to my VP of Operations who expressed concerns about our AI pilot timeline in yesterday's meeting"

2. **Omitting context:** Not telling AI about your organization, team dynamics, or stakeholder's situation to provide background and foundation

3. **Over-engineering:** Writing unnecessary details in prompts or reusing the same, long conversation with many unrelated topics can confuse the AI

4. **Forgetting your audience:** Not specifying who the content is for, their knowledge level or clearly articulating what action you would want them to take

5. **Inconsistent instructions:** Asking for contradictory outputs like "comprehensive but brief"

6. **Not iterating:** Accepting the first response instead of refining and improving it

The best sales professionals treat prompting as a skill to develop, not just a task to complete. Avoiding these mistakes will help you excel at the first 20% of the 20-60-20 framework, setting you up for better results with less editing time.

Employees who say their leaders provide clear AI guidance are 5× more likely to use AI effectively in their daily work.

AI adoption doesn't rise with tools; it rises with leadership. Clear expectations are the difference between experimentation and real performance gains.
(Source: McKinsey)

Essential AI Tools for Sales

When the tool fits the task, the results come fast

Your Perfect AI Partner

Not all AI tools are created equal, and the sales pros and leaders who see the biggest impact are not the ones chasing every new trend. They choose tools with intention and stay focused on what supports their goals, fits their workflows, and removes friction from their selling process.

The best place to start is with the ecosystem you already rely on. If your company uses Microsoft, for example, Copilot will usually outperform a stand-alone app because it works directly inside your daily tools. The same principle applies everywhere: the closer the AI is to your current workflow, the bigger the impact.

You will still need specialty tools for specialty work. Use generalists for breadth and specialists for craft. ChatGPT can cover a lot of ground, but dedicated tools deliver stronger results when quality or depth matters.

It also matters how each tool treats your data. Sales pros and leaders should understand where information is stored, how it is protected, and what policies govern its use. Data practices are not fine print. They are part of the decision.

Experiment and explore, but do not disappear into the endless tool rabbit hole. The best AI tool is the one you and your team will use consistently because it makes your work easier, faster, or smarter.

This chapter shares the tools we rely on and the ones our clients and colleagues continually rave about.

Pick The Tools You'll Use

Select the Right Apps

Leaders and teams don't need vast toolkits. You need the right tools; the ones that add clarity, speed, and consistency to the work you already do (or enable new, valuable capabilities). To choose wisely, look for signals that a tool will actually make a positive difference:

1. **Does it solve a real problem?**
   Tools should deliver ROI. Will it reduce effort, improve quality, or speed things up? If you can't prove value, don't add it.

2. **Will your team see immediate value?**
   Early wins matter. Choose tools that produce visible results fast so adoption feels rewarding instead of burdensome.

3. **Does it integrate smoothly?**
   A great tool that sits off to the side will get ignored. Favor options that fit naturally into daily routines and minimize context switching.

4. **Can you support it?**
   Pick tools that your team can learn, troubleshoot, and maintain without turning every question into a help-desk ticket.

Bottom line: The best tool is the one your team adopts without hand-holding because it clearly improves their work from day one.

Be AI-Safe

Understand the Risk Before You Pick

TIP #
24

AI security is not about being fearful. It is about being practical. Treat AI tools with the same discipline you'd apply to any system that handles information and supports decision-making.

1. **Know where your data goes.**
   Every tool has its own rules for storage, retention, and training. A quick check of the settings or documentation can prevent long-term headaches.

2. **Use the right environment for the right work.**
   Public models are great for brainstorming and drafts. Internal or enterprise-approved tools are better for anything sensitive or strategic.

3. **Watch for accidental oversharing.**
   People often paste more than they realize into AI tools. Encourage teams to pause before dropping in full emails, customer details, or internal documents.

4. **Review outputs with a critical eye.**
   Even when security is solid, accuracy varies. Treat AI-generated content as a draft, not a final decision or official record.

When you model responsible use that's explainable and thoughtful, colleagues follow suit. This is even more if you're a leader! Good judgment is one of your most important, irreplaceable human skills.

ChatGPT

TOOL #
25

Your Strategic Thinking Partner

ChatGPT is fast, versatile, and handles pretty much anything you throw at it. The free version works great for most tasks, but you should upgrade to ChatGPT Plus (approximately $20/month) for faster responses and better reasoning. This is what we use every day: drafting prospecting emails, prepping for discovery calls, brainstorming objection responses, and getting quick answers to complicated questions. The "Canvas" feature is super helpful for editing longer documents like proposals, and Projects helps us stay organized by account or deal. Just double-check any stats or company details before you share them.

https://chat.openai.com

General Purpose

✳ Claude

TOOL #
26

The Detail-Oriented Analyst

Claude is where we go for the heavy lifting. It handles nuance and context better than most, especially when you need to analyze long documents, compare competing proposals, or write content that actually sounds like you. This is our pick for RFP responses, detailed account research, and anything that needs a more polished touch. We recommend the paid plan for serious use. If you are working on complex deals with lots of stakeholders and documentation, Claude handles that depth better than most AI tools out there.

https://claude.ai

Writing & Coding

Copilot

Microsoft 365 Powerhouse

If you use Microsoft Office, start here. Copilot works directly inside Word, Excel, PowerPoint, and Outlook, so you are not copying and pasting between tools. It can draft follow-up emails, analyze pipeline data in spreadsheets, build presentations for quarterly business reviews (QBRs), and summarize meeting notes without leaving your workflow. Enterprise-grade security and data protection make it the go-to choice for companies serious about keeping customer information locked down. Works best when your organization is already in the Microsoft ecosystem.

https://copilot.microsoft.com

Workplace Integration

Gemini

Google Workspace Partner

Gemini integrates beautifully with Google Workspace. Ask it to analyze your Gmail inbox for follow-up opportunities, summarize Google Docs from client meetings, or pull insights from your Drive files without switching platforms. It handles text, images, code, and data across different formats, making it surprisingly versatile for prospect research and creative problem-solving. Best for sales pros already living in Google's ecosystem who want AI that works seamlessly with their existing tools. The integration alone saves significant time on daily administrative work.

https://claude.ai

Research & Multimodal

perplexity

TOOL # 29

Research Assistant

Perplexity is a web-connected AI that is great for quick prospect research, current company information, and getting rapid briefings before calls. It cites its sources, making it easier to verify information before you share it with a prospect. Think of it as a smarter search engine that gives you actual answers instead of just links. Perfect for competitive intelligence, industry trends, or when you need up-to-date facts fast. Just remember it can occasionally get details wrong, so verify anything critical before putting it in a proposal.

https://www.perplexity.ai

Research

NotebookLM

TOOL # 30

Capture, Research, Synthesize

NotebookLM turns your documents, slides, PDFs, videos, and websites into a searchable workspace that is easy to interrogate and reuse. Upload a prospect's annual report, their recent press releases, and your discovery call notes, then ask questions across all of it at once. Its standout features are infographics and an audio overview (turns dense material into a podcast-style conversation). The free plan works well for most sales pros. We strongly recommend it for deep account research before important meetings or as a notebook for past RFP responses for quick reference.

http://notebooklm.google.com

Research & Learning

JULIE HOLMES
SMARTER HQ

✳ Apollo

Research Assistant

Apollo combines a massive B2B database with AI-powered engagement tools, helping you identify ideal prospects and connect with them through automated, personalized sequences. Search by company size, industry, tech stack, funding status, and dozens of other filters to build targeted lists in minutes. The platform handles everything from finding prospects to engaging them across email, phone, and LinkedIn. For sales pros who need prospecting and outreach in one place without juggling multiple tools, Apollo delivers serious value. The free tier is generous enough to test whether it fits your workflow.

https://apollo.io

Prospecting & Outreach

🌈 clay

Data Enrichment Powerhouse

Clay connects 75+ data enrichment sources and AI-powered messaging into one platform, letting you build hyper-personalized outreach at scale. Pull company news, job changes, tech stack details, and funding data automatically, then use AI to turn those signals into relevant opening lines. It is like having a research team that works while you sleep. The learning curve is steeper than simpler tools, but the payoff is outreach that actually feels personal. Best for sales pros and teams who are serious about standing out in crowded inboxes.

http://clay.com

Research & Learning

Lusha

TOOL #
33

Contact Data Platform

Lusha helps sales teams find and verify direct dials and email addresses so you can actually reach decision-makers. The browser extension works while you browse LinkedIn or company websites, surfacing contact details in real time. Accuracy rates are solid, and the platform integrates with most major CRMs for easy syncing. For sales pros tired of bounced emails and generic info@ addresses, Lusha removes a major friction point. The free tier gives you enough credits to test it out before committing.

https://lusha.com

Contact Intelligence

6sense

TOOL #
34

Account Engagement Platform

6Sense uses AI to identify accounts that are actively researching your product category, even if they have never visited your website. It analyzes billions of buying signals across the web to predict which accounts are in-market and where they are in the buying journey. This means you can prioritize outreach to prospects already looking for solutions like yours. It is an enterprise platform with enterprise pricing, but worth knowing about if your organization is ready to invest in intent data. The insights can dramatically improve how you allocate your selling time.

http://6sense.com

Intent Data & ABM

GONG

Conversation Intelligence Platform

Gong is the gold standard for analyzing sales conversations across calls, emails, and video meetings. It identifies what top performers do differently, tracks customer sentiment, spots competitive mentions, and highlights risks and opportunities in your pipeline. The coaching insights show you exactly where deals are strong or stalling. For sales leaders, it is a window into what is actually happening on calls without listening to hours of recordings. For reps, it is like having a coach review every conversation and point out what to do differently next time.

https://gong.io

Conversational Intel

FATHOM

AI Meeting Assistant

Fathom records, transcribes, and summarizes your sales calls across Zoom, Microsoft Teams, and Google Meet without you lifting a finger. You can ask questions about your transcript, adjust summaries to fit your needs, and share meeting notes with one click. The time savings are massive, especially when you are back-to-back all day and still need to update your CRM. The free plan is genuinely useful and works great for individuals or small teams. Best for anyone tired of taking notes during important prospect conversations.

https://fathom.video

Meeting Intelligence

RingCentral®

TOOL # 37

AI-Powered Comms Platform

RingCentral's RingSense AI works across calls, video meetings, and messages to capture insights you would otherwise miss. It transcribes conversations in real time, highlights action items, and analyzes sentiment to help you read the room even when you are juggling multiple deals. The AI coach provides feedback on communication patterns, helping you improve how you handle objections and discovery questions. It integrates smoothly with existing workflows and CRMs, making it particularly valuable for sales teams managing high call volumes.

https://ringcentral.com

Meetings & Collaboration

• •

SUPERHUMAN

TOOL # 38

Speed-Focused Email Client

Superhuman is for sales pros obsessed with inbox zero and brutal efficiency. It replaces Gmail or Outlook with a lightning-fast interface built around keyboard shortcuts, smart filtering, and AI-powered email drafts. The AI helps you write faster and suggests follow-ups based on your email patterns. Starting at $30/month, it is expensive but transformative if email volume is crushing your selling time. The onboarding includes personal training to build new habits. Best for high-volume sellers who live in their inbox and want to claw back hours every week.

http://superhuman.com/

Email Efficiency

LAVENDER

TOOL 39

AI Email Coach

Lavender analyzes your sales emails before you send them, providing specific suggestions to improve response rates. It evaluates subject lines, email length, tone, readability, and call-to-action effectiveness based on data from millions of sales emails. You get a score and actionable fixes in real time as you write. It integrates directly into Gmail, Outlook, and most sales engagement platforms. For sales pros who want to improve their cold outreach without guessing what works, Lavender gives you the feedback loop most people never get. The free tier lets you test it before committing.

https://lavender.ai

Email Coaching

grammarly

TOOL 40

Writing Quality Guardian

Grammarly runs quietly in the background across all your platforms, catching spelling and grammar errors in real time. We keep it running so everything we write gets a quick polish before it goes out. The free version handles basics well, but Grammarly Pro adds tone suggestions, style improvements, and helps you write clearer, more professional communications. For sales pros sending dozens of emails and messages daily, it prevents the small mistakes that undermine credibility. Best for anyone who wants error-free communication without thinking about it.

http://grammarly.com

Writing & Editing

instantly

Speed-Focused Email Client

TOOL #
41

Instantly helps you send cold outreach across multiple email accounts without landing in spam. It handles email warm-up automatically, rotates sending accounts, and tracks deliverability so your messages actually reach prospects. The AI helps you write and optimize sequences based on what is getting replies. For sales pros and teams doing high-volume prospecting, it solves the technical headaches that kill cold email campaigns. The interface is clean and the pricing is accessible compared to enterprise alternatives. Best for SDRs and account executives who need to fill pipeline fast without deliverability disasters.

http://instantly.ai

Cold Email Automation

Outreach

Sales Execution Platform

TOOL #
42

Outreach is the platform many enterprise sales teams use to execute multi-channel sequences with AI-powered optimization. It coordinates email, phone, LinkedIn, and other touchpoints into structured workflows that keep deals moving. The AI suggests the best times to reach out, recommends next actions, and flags deals that need attention. For sales organizations that want consistency and visibility across their entire revenue process, Outreach delivers the structure. It is an enterprise tool with enterprise pricing, but worth knowing if your team is scaling beyond spreadsheets and basic CRMs.

https://outreach.io

Sales Engagement

Salesloft.

Revenue Workflow Platform

Salesloft offers a comprehensive platform for executing sales cadences with AI-powered coaching built in. It helps you build multi-touch sequences, provides real-time guidance during calls, and suggests the best content for each prospect based on where they are in the buying journey. The analytics show what is working across your outreach so you can double down on winning patterns. Like Outreach, it is built for teams that need structure and scale. If your organization uses Salesloft, mastering its AI features will make your sequences significantly more effective.

http://salesloft.com

Sales Engagement

● CRAYON

AI-Driven Competitive Analysis

Crayon automatically tracks your competitors across their websites, pricing changes, messaging, product updates, and market moves, then alerts you to what actually matters. It turns scattered competitive intel into organized insights your sales team can use through battlecards integrated into Salesforce and Slack. For sales pros going up against the same competitors repeatedly, having real-time intelligence changes how confidently you handle comparison questions. Best for teams that want to win more competitive deals without manual research.

https://crayon.co

Competitive Intel

JULIE HOLMES
SMARTER HQ

HeyGen

AI Avatar and Video Translation

HeyGen creates professional videos using AI avatars that look and sound remarkably real. You can create a digital twin that mirrors your expressions and gestures, so you can produce personalized prospecting videos without recording each one individually. The translation feature converts videos into 175+ languages with perfect lip-sync, making it look like you are actually speaking that language. For sales pros targeting global accounts or wanting to scale video outreach without the production time, HeyGen delivers. Free plan lets you test it; paid plans start around $24/month for serious use.

https://heygen.com

Video Creation

vidyard®

Video Messaging for Sales

Vidyard is the go-to platform for sales video messaging. Record quick personalized videos from your browser, embed them in emails, and track exactly who watched and for how long. The analytics show you which prospects are engaged so you know who to prioritize for follow-up. It integrates with most CRMs and sales engagement platforms, making video a natural part of your workflow rather than an extra step. For sales pros who want to stand out in crowded inboxes with a personal touch, Vidyard makes video outreach simple and measurable.

http://vidyard.com

Video Messaging

Clari

Revenue Intelligence Platform

Clari gives sales leaders and reps a clearer picture of pipeline health, deal momentum, and forecast accuracy. It analyzes activity data from email, calendar, and CRM to show which deals are progressing and which are stalling, often before you realize it yourself. The AI highlights risks and recommends where to focus your attention. For individual reps, it helps you prioritize your day. For managers, it replaces gut-feel forecasting with data. It is an enterprise platform, but worth knowing about if your organization is serious about predictable revenue.

https://clari.com

Revenue Intelligence

Crystal

Lightning-Fast Transcription

Crystal analyzes prospects' communication styles using DISC personality profiles, helping you tailor your outreach and conversations to how they prefer to receive information. The browser extension shows personality insights while you browse LinkedIn, so you know before a call whether someone wants data and details or prefers big-picture thinking. It suggests how to phrase emails, which words to use, and what approach will resonate. For sales pros tired of guessing why some messages land and others fall flat, Crystal takes the mystery out of personalization.

https://crystalknows.com

Buyer Intelligence

Canva

AI Design & Content Creation

Canva handles everything from one-pagers and proposal graphics to social posts and presentation polish. The AI features generate layouts from prompts, write copy in your brand voice, and reformat designs across platforms instantly. Built-in tools handle background removal, image expansion, and object manipulation without needing design skills. For sales pros creating leave-behinds, custom visuals for proposals, or LinkedIn content, Canva makes you look more polished than you probably are. The free version works, but Pro unlocks the full AI toolkit if you are creating content regularly.

https://canva.com

Design & Marketing

TOOL #
49

- -

GAMMA

Quick Slide & Site Creator

Gamma transforms prompts, pasted text, or uploaded files into clean presentations fast. Describe what you need, and it generates structure, suggests visuals, and handles formatting automatically. It is perfect for pitch decks, QBR presentations, or turning discovery notes into a visual leave-behind for your champion. You can generate complete presentations in minutes that actually look polished. For sales pros who need decks frequently but hate building slides from scratch, Gamma removes the friction. The free tier is generous enough to handle most individual needs.

https://gamma.app

Presentations

TOOL #
50

Companies where the CEO oversees AI governance report significantly higher financial impacts from their AI investments

Leadership commitment matters. When executives champion AI adoption in sales, teams see faster implementation and better results than when AI is treated as just another tech project.
(Source: McKinsey & Company)

ChatGPT Prompts for Success

Sell Smarter and Close Faster with These Ready-to-Use AI Prompts

How to Use These Prompts

Here's a familiar scenario: You're juggling a full pipeline, back-to-back calls, and still drowning in follow-ups that should have been sent hours ago. That's where these prompts come in.

These AI prompts are your starting point, not rigid scripts. Think of them as flexible frameworks you can adapt to fit your selling style, prospects, and deal situations. Here's how to get the most out of them:

1. **Fill in the Blanks**: Add your specific context so the AI delivers insights that actually move the deal forward. The more detail you give, the better the output.

2. **Make Them Your Own**: Adjust the language, weave in examples from your industry, or combine prompts. Top performers don't copy-paste, they customize for impact.

3. **Experiment and Refine**: If the first answer isn't quite it, tell the AI what to tweak: "Make this more conversational," "Focus on their budget concerns," or "Highlight ROI." You'll dial it in fast.

4. **Save What Works**: Build your personal library of "power prompts." This becomes your AI sales playbook, and it'll save you hours every single week (yes, really).

Create an ROI Calculator

Sales Strategy & Messaging

Prompt

You are a financial analyst who specializes in building compelling business cases for B2B solutions. Your job is to help me create an ROI calculator framework that I can customize for individual prospect conversations.

Start by asking up to five clarifying questions so you understand my product, pricing model, and the metrics that matter most to my buyers. Then create a calculator structure that includes: current state costs and inefficiencies the prospect should quantify, implementation considerations and timeline to value, projected benefits over one to three years covering both hard dollar savings and strategic advantages, and a simple summary view that makes the business case clear to financial decision-makers.

For each section, suggest the specific questions I should ask prospects to gather accurate inputs, explain how to present the numbers persuasively without overpromising, and flag where buyers typically push back on assumptions. The goal is a practical tool I can adapt for different deal sizes and industries, not an overly complex spreadsheet. Keep calculations straightforward and defensible.

Use the information below to personalize your response:
– Product/service name and primary function:
– Key cost metrics it impacts for customers:
– Average implementation timeframe:
– Typical deal size range:
– Primary value drivers (list 3-5 benefits):
– Common financial objections you face:

Competitor Battlecard

Sales Tactics

Prompt

You are a competitive intelligence analyst who helps sales teams win more deals against specific competitors. Your job is to create a practical, single-page battlecard I can reference during or before competitive sales conversations.

Start by asking up to five clarifying questions so you understand my solution, the competitor, and typical deal scenarios where we go head-to-head. Then build a battlecard that includes: a brief competitor overview and how they typically position themselves, a comparison highlighting where we have clear advantages, specific weaknesses in their offering I can probe for during discovery, effective responses to the claims they commonly make, and questions I can ask prospects that naturally favor our strengths.

Organize everything in a scannable format that works mid-call. Focus on substantive differences that actually influence buying decisions, not minor feature comparisons. Keep all claims factual and defensible.

Use the information below to personalize your response:
– Your product/service category:
– Competitor name:
– Your top 3-5 differentiating strengths:
– Competitor's main selling points:
– Competitor's known weaknesses:
– Typical buyer concerns in competitive deals:

BANT Qualification List

Sales Tactics

Prompt

You are an experienced sales consultant who specializes in pipeline quality and deal qualification. Your job is to create a practical BANT (Budget, Authority, Need, Timeline) qualification checklist I can use during discovery calls to quickly assess whether an opportunity is worth pursuing.

Start by asking up to five clarifying questions so you understand my solution, typical buyers, and what a qualified deal looks like for my business. Then build a comprehensive checklist that includes: three to five specific questions for each BANT element that feel conversational rather than interrogative, signals and responses that indicate a strong qualified opportunity, and red flags or vague answers that suggest the deal may stall or not be real. Add a dedicated section on understanding their internal decision-making process, buying committee dynamics, and competing priorities, since these factors often derail otherwise well-qualified deals.

Include guidance on how to gracefully disqualify opportunities that are not a fit without burning the relationship for future outreach. Format this as a practical reference I can glance at during live conversations without losing my place. The goal is helping me invest time in opportunities that will actually close and politely move on from those that will not.

Use the information below to personalize your response:
– Your product/service:
– Typical deal size:
– Average sales cycle length:
– Common buyer titles/roles:
– Top reasons deals stall or fall through:

JULIE HOLMES
SMARTER HQ

Price Negotiation Talking Points

Sales Tactics

PROMPT #
54

Prompt

You are an expert in B2B sales negotiation who helps sales professionals handle pricing conversations with confidence. Your job is to create a set of talking points I can use when prospects push back on price, ask for discounts, or compare us unfavorably to cheaper alternatives.

Start by asking up to five clarifying questions so you understand my solution, pricing structure, and the value metrics that justify our price point. Then create talking points that include: language that acknowledges budget concerns without immediately conceding, specific phrases and frameworks for shifting conversations from price to total value and ROI, confident responses to common objections like "your competitor is 30% cheaper" and "we need a discount to get this approved internally," techniques for understanding what is really driving the price pushback before responding, and approaches for holding firm on pricing without damaging the relationship or sounding inflexible.

Include guidance on when it might make sense to offer concessions and how to do so strategically without setting a bad precedent. The tone should be confident and collaborative, not defensive or aggressive. I want to protect my margins while still making the prospect feel heard and respected.

Use the information below to personalize your response:
– Product/service and pricing structure:
– Key differentiators that justify your pricing:
– Common price objections you hear:
– Value metrics or ROI data you can reference:
– Typical discount authority or thresholds:
– Competitors prospects usually compare you to:

Objection-Handling Cheat Sheet

Sales Tactics

Prompt

You are a sales strategist who specializes in helping reps handle objections smoothly and confidently. Your job is to create a comprehensive objection-handling cheat sheet I can reference before and during sales conversations. Start by asking up to five clarifying questions so you understand my solution, typical buyer concerns, and where deals usually get stuck. Then build a cheat sheet covering the most common objections I face, with two to three effective response options for each.

Structure every response to first acknowledge the concern genuinely so the prospect feels heard, then reframe the conversation toward value or a different perspective, and finally guide toward a logical next step that keeps the deal moving. Include objections related to price and budget, timing and urgency, competition and alternatives, need for internal approval or consensus, and satisfaction with their current solution or process.

For each objection, note what the prospect is really saying underneath the surface concern, since that context helps me respond more effectively. Keep the language conversational and natural rather than scripted and salesy. The goal is building trust and momentum, not winning an argument.

Use the information below to personalize your response:
– Your product/service:
– Most common objections (list 5-7):
– Your key differentiators:
– Typical buyer persona and seniority level:
– Preferred tone (consultative, direct, casual, etc.):

JULIE HOLMES
SMARTER HQ

Incumbent Vendor Rebuttal

Sales Tactics

PROMPT #
56

Prompt

You are an expert sales strategist who helps reps break through the toughest early-conversation objection: "We're happy with our current vendor." Your job is to create a rebuttal approach that opens doors without being pushy or disrespectful of their existing relationship.

Start by asking up to five clarifying questions so you understand my solution, how it differs from incumbent vendors, and what typically causes companies to eventually switch. Then create a response framework that includes: two to three rebuttal variations ranging from soft and curious to more direct and challenging, specific questions I can ask to uncover hidden dissatisfaction or unmet needs without sounding like I am criticizing their current choice, language that positions a conversation with me as low-risk and educational rather than a hard sell, techniques for planting seeds about gaps they may not realize they have, and a graceful exit approach that keeps the door open for future outreach if they genuinely are not ready today.

The goal is creating a bridge to continue the dialogue and earn a real conversation, not pressuring them into an immediate switch. Help me be respectfully persistent without crossing into annoying.

Use the information below to personalize your response:
– Your product/service category:
– How you typically differ from incumbent solutions:
– Pain points that existing vendors often miss:
– Triggers that cause companies to eventually switch:
– Your ideal low-commitment next step (audit, assessment, demo, etc.):

Multi-Stakeholder Sales Framework

Complex Sales Strategies

Prompt

You are a strategic sales consultant who specializes in complex B2B deals involving multiple decision-makers. Your job is to help me create a framework for navigating buying committees and building consensus across stakeholders with different priorities. Start by asking up to five clarifying questions so you understand my solution, typical deal complexity, and the roles usually involved in purchase decisions. Then build a comprehensive framework that includes: strategies for mapping the full buying committee early and identifying stakeholders I might not have met yet, techniques for understanding each person's unique priorities and what success looks like from their perspective, approaches for tailoring my value proposition to resonate with different roles like finance, operations, IT, and executive sponsors, methods for building internal champions and equipping them to sell on my behalf when I am not in the room, tactics for building consensus when stakeholders have competing priorities or concerns, and warning signs that a stakeholder may be blocking the deal along with approaches for addressing resistance. Include guidance on maintaining momentum when different committee members have varying levels of engagement and responsiveness. The goal is turning complex multi-threaded deals from chaotic to systematic so I can navigate them with confidence rather than hope.

Use the information below to personalize your response:
– Your product/service category:
– Typical buying committee size and roles:
– Average deal size and sales cycle length:
– Primary value drivers for different stakeholder types:
– Common objections or concerns from different roles:
– Where deals typically stall in the committee process:

Custom Demo Script

Sales Tactics

Prompt

You are an experienced sales consultant who helps reps deliver demos that feel consultative rather than like a feature tour. Your job is to create a tailored demo script for a specific prospect that keeps them engaged and moves toward clear next steps.

Start by asking up to five clarifying questions so you understand the prospect's situation, what they care about most, and what a successful demo would accomplish. Then build a demo script that includes: a strong opening that confirms their priorities and sets an agenda focused on their needs rather than my features, a logical flow that shows only the capabilities relevant to their stated challenges and skips everything else, discovery questions woven throughout to maintain dialogue and uncover new information rather than presenting in monologue, specific moments to pause and check for reactions or concerns, techniques for handling objections or skeptical questions that come up live without getting derailed, and a clear closing that summarizes value and proposes a specific next step. Include guidance on pacing and how to adjust on the fly if they want to dive deeper on something or move faster. The goal is a demo that feels like a conversation about solving their problems, not a product walkthrough.

Use the information below to personalize your response:
– Prospect's company, industry, and size:
– Their key challenges or needs (from discovery):
– Features most relevant to their situation:
– Stakeholders attending the demo and their roles:
– Common objections or questions during demos:
– Your ideal next step after the demo:
– Time allocated for the demo:

Craft Email Subject Lines

Prospecting & Outreach

PROMPT #
59

Prompt

You are an expert email marketer who specializes in B2B sales outreach and knows what makes prospects actually open cold emails. Your job is to generate a diverse set of compelling subject lines I can test across my prospecting campaigns.

Start by asking up to five clarifying questions so you understand my target audience, what I am offering, and what has or has not worked in the past. Then create fifteen subject lines using a variety of approaches: curiosity-driven lines that make them want to know more, value-proposition lines that promise a specific benefit, personalization-focused lines that reference something specific about them or their company, problem-focused lines that call out a challenge they likely face, and social proof lines that reference similar companies or results.

Keep every subject line under sixty characters so it displays fully on mobile, avoid spam trigger words and anything that feels clickbait or gimmicky, and make sure each one sounds like it came from a real person rather than a marketing automation tool. For each subject line, include a brief note on when it works best and what type of prospect it would resonate with most. The goal is a tested toolkit I can rotate through rather than using the same tired approaches.

Use the information below to personalize your response:
– Your target buyer persona and title:
– Industry or vertical you are targeting:
– Your primary value proposition:
– Common pain points your prospects face:
– Any subject lines that have worked well before:
– Any approaches you want to avoid:

Customized Cold Email

Prospecting & Outreach

Prompt

You are an expert in B2B sales outreach who specializes in writing cold emails that actually get responses. Your job is to craft a personalized cold email for a specific prospect that feels relevant and human rather than templated and mass-produced.

Start by asking up to five clarifying questions so you understand the prospect, their likely challenges, and what would make them want to respond. Then write an email that includes: an opening line that references something specific about them or their company so they know this is not a blast, a brief and clear explanation of why you are reaching out and what problem you help solve, a credibility element like a relevant result or recognizable customer that builds trust without bragging, and a low-friction call to action that makes responding easy.

Keep the entire email under 150 words since busy executives will not read anything longer. Avoid generic phrases like "I hope this email finds you well" or "I wanted to reach out" that signal automation. The tone should be professional but conversational, like a smart colleague sending a helpful note rather than a salesperson pushing a pitch. After the draft, explain why you made the choices you did so I can learn from the approach.

Use the information below to personalize your response:
– Prospect's name, title, and company:
– Something specific about them or their company (news, LinkedIn post, initiative):
– Your product/service and primary value:
– A relevant result or customer you can reference:
– Your ideal call to action (reply, call, meeting, etc.):

Industry-Specific Prospecting Emails

Prospecting & Outreach

Prompt

You are a B2B sales messaging strategist with deep expertise across multiple industries. Your job is to create prospecting email templates tailored to specific verticals that demonstrate I understand their world rather than sending generic outreach.

Start by asking up to five clarifying questions so you understand my solution, how value differs by industry, and which verticals I am prioritizing. Then create five industry-specific prospecting templates, one for each vertical I specify. Each template should be 125 to 175 words and include: an opening that references an industry-specific challenge or trend they will immediately recognize, language and terminology that sounds like I work in their space, a value statement framed around outcomes that matter in their industry rather than generic benefits, a relevant proof point or example from a similar company if available, and an appropriate call to action for their typical buying behavior.

Note which parts of each template should be personalized for individual prospects versus used as written. Also flag any compliance or sensitivity considerations for regulated industries. The goal is outreach that makes prospects think "this person actually understands my business" rather than "this is another vendor who found my email."

Use the information below to personalize your response:
– Your product/service:
– Five industries or verticals to create templates for:
– How your value proposition differs by industry:
– Example customers in each vertical (if available):
– Key pain points you address in each vertical:
– Your typical call to action:

Sales Video Scripts

Prospecting & Outreach

Prompt

You are a video marketing strategist who specializes in creating high-impact sales videos that get watched and drive responses. Your job is to develop three video script templates at different lengths (30 seconds, 60 seconds, and 90 seconds) that I can customize for prospecting, follow-ups, and proposal walk-throughs.

Start by asking up to five clarifying questions so you understand my audience, what I am selling, and how I plan to use video in my sales process. Then create three scripts that each include: a strong opening hook in the first five seconds that earns their attention before they click away, a structured middle section that delivers value and builds credibility without rambling, and a clear call to action that tells them exactly what to do next. Include specific guidance on where to personalize each script for individual prospects, recommendations on tone and energy level appropriate for each length and purpose, and tips on what to show visually or have in the background to reinforce credibility.

The 30-second version should work for cold outreach, the 60-second for warm follow-ups, and the 90-second for proposal introductions or champion enablement. Keep language conversational and natural rather than scripted and stiff.

Use the information below to personalize your response: – Your product/service and primary value:
– Target buyer persona and typical seniority:
– Common pain points you address:
– Your typical next step after someone watches:
– Your personal communication style (casual, professional, energetic, calm):
– Where you plan to send these videos (email, LinkedIn, etc.):

Long Sales Cycle Touchpoints

PROMPT #

63

Prospecting & Outreach

Prompt

You are a sales strategist who specializes in managing complex deals with extended sales cycles. Your job is to create a comprehensive touchpoint sequence that keeps me relevant and top-of-mind with prospects over months without being annoying or repetitive.

Start by asking up to five clarifying questions so you understand my typical sales cycle length, buyer journey stages, and what kind of content and interactions resonate with my prospects. Then build a multi-month sequence that includes: a mix of channels including email, phone, LinkedIn, and video spread strategically across the timeline, specific touchpoint goals for each stage from initial awareness through decision, content recommendations for each touch that provide genuine value rather than just "checking in," guidance on spacing and frequency that maintains presence without overwhelming, techniques for re-engaging when prospects go quiet at different stages, and triggers or signals that indicate when to accelerate or slow down the cadence. Include at least one creative or unexpected touchpoint that helps me stand out from competitors running standard sequences. The goal is a systematic approach that builds relationship and credibility over time so that when they are ready to buy, I am the obvious choice.

Use the information below to personalize your response:
– Your product/service:
– Average sales cycle length:
– Typical buyer journey stages:
– Target buyer persona and seniority:
– Content or resources you have available to share:
– Channels that work best with your audience:

Target Account Outreach Plan

Prospecting & Outreach

PROMPT #
64

Prompt

You are an account-based sales strategist who helps reps break into their most important target accounts. Your job is to create a detailed, personalized outreach plan for a specific high-priority prospect that goes beyond standard sequences.

Start by asking up to five clarifying questions so you understand the account, what you know about them, any existing relationships, and why they are a priority target. Then build a comprehensive outreach plan that includes: research tasks to complete before any outreach to find personalization angles and potential triggers, a multi-threaded approach that identifies multiple stakeholders to engage rather than relying on a single contact, a sequenced series of touchpoints across email, phone, LinkedIn, video, and any creative channels appropriate for this account, specific messaging angles tailored to what you know about their business priorities and challenges, techniques for leveraging any mutual connections, customers, or events to warm up the outreach, and contingency approaches if primary contacts do not respond. Include timing recommendations and a realistic assessment of how long this plan should run before adjusting strategy. The goal is treating this account like a campaign rather than a series of random touches.

Use the information below to personalize your response:
– Target company name and industry:
– Why this account is high-priority:
– What you know about their business priorities or challenges:
– Stakeholders or contacts you have identified:
– Any mutual connections, shared customers, or warm angles:
– Previous outreach attempts (if any):
– Resources or content you can leverage:

Voicemail Follow-Up

Prospecting & Outreach

PROMPT #
65

Prompt

You are a sales communication specialist who helps reps maximize response rates from cold outreach. Your job is to create a follow-up email sequence to send after leaving a prospect a voicemail, creating a coordinated multi-touch approach that increases the chance of a response.

Start by asking up to five clarifying questions so you understand the context of your outreach, what you said in the voicemail, and what response you are hoping for. Then create a three-email sequence that includes: an immediate follow-up email sent within minutes of the voicemail that references the message you left and reinforces your reason for reaching out, a second follow-up two to three days later that adds new value or a different angle rather than just repeating yourself, and a third breakup-style email five to seven days later that creates gentle urgency while leaving the door open. Each email should be concise and scannable, reference the voicemail naturally without being awkward about it, and include a clear low-friction call to action. Include guidance on subject lines for each email in the sequence and how to adapt the approach if they listen to the voicemail versus if they do not. The goal is surrounding them with a coordinated message rather than random isolated touches.

Use the information below to personalize your response:
– Your product/service:
– Target buyer persona and title:
– What you typically say in your voicemail:
– Your primary value proposition or hook:
– Your ideal call to action:
– Any personalization details about this prospect:

Reconnect with Last Year's Leads

Prospecting & Outreach

Prompt

You are a sales re-engagement specialist who helps reps restart conversations with prospects who showed interest but never converted. Your job is to create an email sequence for reconnecting with leads from the previous year in a way that feels fresh and valuable rather than desperate or guilt-inducing.

Start by asking up to five clarifying questions so you understand why these leads went cold, what has changed since you last spoke, and what new value you can offer. Then create a three-email sequence that includes: an opening email that acknowledges the time gap naturally without apologizing, references something new or relevant that gives you a reason to reach out now, and offers genuine value rather than just asking for their time again. The second email should take a different angle, perhaps sharing a relevant result, industry insight, or resource that might reignite interest. The third email should be a respectful closing that creates gentle urgency while making it easy for them to say "not now" without guilt. Each email should be under 150 words, avoid phrases like "circling back" or "just checking in" that signal lazy outreach, and acknowledge that their priorities may have shifted. Include subject line recommendations for each email and guidance on spacing. The goal is earning a fresh look, not guilting them into a response.

Use the information below to personalize your response:
– Your product/service:
– Why leads typically went cold (timing, budget, priority, etc.):
– What has changed since last year (new features, results, pricing, etc.):
– Any valuable content or insights you can share:
– Your ideal call to action:
– Typical lead source (inbound, event, referral, etc.):

Discovery Call Questions

Discovery & Research

PROMPT #
67

Prompt

You are a sales strategist who specializes in consultative discovery and helping reps uncover the information that actually wins deals. Your job is to create a set of powerful discovery questions tailored to my specific product and buyer that go beyond surface-level fact-finding.

Start by asking up to five clarifying questions so you understand my solution, typical buyer challenges, and what information I need to qualify and advance opportunities. Then create a structured set of fifteen to twenty discovery questions organized into logical categories: current state questions that understand how they operate today, pain and impact questions that uncover problems and quantify the cost of inaction, decision process questions that reveal how they buy and who is involved, future state questions that help them articulate what success looks like, and urgency questions that surface timeline drivers and competing priorities. Each question should be open-ended and conversational rather than interrogative, designed to get prospects talking rather than giving one-word answers. Include follow-up probes for key questions that help me dig deeper when I hear interesting answers. Flag which questions are essential for qualification versus helpful for positioning. The goal is a discovery framework that uncovers real buyer motivation, not just a checklist of facts.

Use the information below to personalize your response:
– Your product/service:
– Target buyer persona and typical title:
– Industry or vertical:
– Key pain points your solution addresses:
– Information you need to qualify an opportunity:
– Common reasons deals stall after discovery:

Qualifying Questions Creator

Discovery & Research

PROMPT #
68

Prompt

You are an experienced sales consultant who helps reps quickly assess whether a prospect is a good fit and understand their specific requirements. Your job is to create a set of qualifying questions that efficiently uncover client needs, priorities, and readiness to buy without feeling like an interrogation.

Start by asking up to five clarifying questions so you understand my solution, ideal customer profile, and what separates good-fit prospects from time-wasters. Then create a structured question set that includes: fit questions that determine whether they match my ideal customer profile, need questions that uncover specific problems and requirements my solution should address, priority questions that reveal how important solving this problem is relative to other initiatives, capability questions that surface any technical, resource, or organizational constraints, and buying process questions that clarify decision-making authority, timeline, and budget reality. For each question, include what a good answer sounds like versus a red flag answer so I can calibrate in real-time. Suggest a logical flow for asking these questions that feels like a natural conversation rather than a qualification checklist. Include guidance on how to gracefully exit or deprioritize conversations when fit is not there. The goal is respecting both my time and theirs by quickly getting to the truth about whether we should keep talking.

Use the information below to personalize your response:
– Your product/service:
– Ideal customer profile (size, industry, situation):
– Key problems you solve:
– Deal-breakers or disqualifying criteria:
– Typical sales cycle and deal size:
– Common time-wasters or bad-fit scenarios:

JULIE HOLMES
SMARTER HQ

Pre-Call Research Brief

Discovery & Research

PROMPT #
69

Prompt

You are a sales research analyst who helps reps walk into calls fully prepared and confident. Your job is to create a comprehensive pre-call research brief for a specific prospect that gives me everything I need to have a relevant, informed conversation.

Start by asking up to five clarifying questions so you understand who I am meeting with, what I already know, and what kind of call this is (cold outreach, discovery, demo, negotiation, etc.). Then compile a research brief that includes: a company snapshot covering what they do, size, recent news, and any public information about priorities or challenges, background on the specific person I am meeting including their role, tenure, LinkedIn highlights, and anything that suggests their communication style or priorities, potential pain points they might be experiencing based on their industry, role, and company situation, connections or commonalities I can reference to build rapport naturally, questions I should ask based on what I do and do not know yet, topics or landmines I should be careful about based on any sensitive news or situations, and suggested talking points that connect my solution to their likely concerns. Organize everything in a scannable format I can review in five minutes before the call. The goal is walking in feeling like I have done my homework without spending an hour on research.

Use the information below to personalize your response:
– Prospect's name and title:
– Company name:
– Type of call (discovery, demo, follow-up, negotiation):
– What you already know about them:
– Your product/service and primary value:
– Any specific angles or topics you want to explore:

JULIE HOLMES
SMARTER HQ

Industry Trend Analysis

Discovery & Research

PROMPT #
70

Prompt

You are a market intelligence analyst who helps sales professionals understand the trends shaping their target industries. Your job is to create a concise, actionable trend analysis I can use to have smarter conversations with prospects and position my solution in context of what is happening in their world.

Start by asking up to five clarifying questions so you understand which industry I am focused on, what my solution does, and how I typically engage with prospects in this space. Then create an analysis that includes: an overview of the three to five most significant trends affecting this industry right now and why they matter, the business drivers behind each trend including economic shifts, technology changes, regulatory pressures, or customer behavior changes, how each trend is impacting companies in this industry and the challenges it creates, specific opportunities where my type of solution becomes more relevant or urgent because of these trends, and talking points or questions I can use in sales conversations to demonstrate I understand their landscape. Include any data points or statistics that would add credibility when referencing these trends. Flag which trends are well-established versus emerging so I know how to frame them. The goal is sounding like someone who genuinely understands their industry rather than a vendor who just wants to sell something.

Use the information below to personalize your response:
– Industry or vertical to analyze:
– Your product/service and primary value:
– Geographic focus (if relevant):
– Target buyer persona:
– Any specific trends you have heard prospects mention:
– How you typically position your solution in this industry:

Analyze Your Competition

Discovery & Research

PROMPT #
71

Prompt

You are a senior strategy consultant and competitive intelligence analyst who helps sales teams understand their competitive landscape. Your job is to help me analyze where we stand against competitors so I can position more effectively and win more deals. Start by asking up to seven clarifying questions so you understand my solution, the competitors I face most often, and what I already know about the competitive dynamics. Then create a comprehensive analysis that includes: an overview of the competitive landscape organized into logical groupings of competitors with brief descriptions of each, a detailed comparison of my solution versus key competitors across dimensions that matter to buyers including capabilities, pricing approach, go-to-market model, and typical customer profile, an honest assessment of where we have clear advantages and where competitors may have an edge, hidden competitors or alternative approaches that prospects might consider instead of a solution like mine, specific positioning strategies for common competitive scenarios including how to lead when we are strong and how to neutralize when we are weaker, and questions I can ask in deals to uncover competitive dynamics early and steer toward our strengths. Be direct about gaps or weaknesses since I need realistic guidance rather than cheerleading. The goal is going into competitive deals with clear strategy rather than hoping our features speak for themselves.

Use the information below to personalize your response:
– Your product/service:
– Industry and target customer profile:
– Main competitors you encounter in deals:
– Your key differentiators and strengths:
– Areas where competitors typically win or have advantages:
– Common competitive objections you hear from prospects:
– Any specific competitive deals you want to analyze:

Summarize Sales Call Key Points

Follow-Up & Relationship

PROMPT #
72

Prompt

You are a sales communications expert who helps reps turn conversations into clear, actionable follow-up. Your job is to take my notes from a sales call and transform them into a professional summary email that reinforces key points, confirms next steps, and keeps the deal moving forward. Start by asking up to five clarifying questions so you understand the context of the call, who was involved, and what outcomes we are working toward. Then create a follow-up email that includes: a brief recap of the main topics discussed organized in a logical and scannable format, clear acknowledgment of the prospect's key concerns, priorities, or requirements they shared, confirmation of any decisions made or agreements reached during the conversation, specific next steps with owners and timing so nothing falls through the cracks, and a professional closing that reinforces your commitment and makes it easy for them to respond with questions or corrections. Keep the email concise and skimmable since busy executives will not read paragraphs. Use bullet points strategically but do not make it feel like a robotic meeting transcript. The tone should feel like a thoughtful colleague summarizing a productive conversation rather than a salesperson documenting a transaction. After the draft, suggest a subject line that will get opened and flag anything I should consider adding based on what I shared.

Use the information below to personalize your response:
– Prospect's name, title, and company:
– Who else was on the call:
– Type of call (discovery, demo, negotiation, etc.):
– Main topics discussed:
– Key concerns or priorities they shared:
– Decisions made or next steps agreed upon:
– Your notes or transcript from the call:

LinkedIn Message After Event

PROMPT #

73

Follow-Up & Relationship

Prompt

You are a networking and social selling expert who helps sales professionals turn brief event encounters into real business conversations. Your job is to create a LinkedIn follow-up message that feels personal and relevant rather than like a generic post-event blast. Start by asking up to five clarifying questions so you understand the event, your interaction with this person, and what kind of relationship you are hoping to build. Then craft a message that includes: an opening that specifically references the event and your conversation so they immediately remember who you are, a genuine observation or compliment about something they shared or presented that shows you were actually paying attention, a brief mention of why you would like to stay connected that focuses on mutual value rather than your sales pitch, and a low-pressure call to action that makes it easy to respond without committing to anything significant. Keep the entire message under 100 words since LinkedIn messages that feel like emails get ignored. Avoid anything that sounds like a pitch or makes them feel like they just got added to a sales sequence. The tone should feel like a friendly professional following up on a good conversation rather than a rep hunting for leads. Suggest a connection request note if we are not already connected and a message if we are.

Use the information below to personalize your response:
— Event name and type (conference, networking, webinar, etc.):
— Prospect's name and role:
— What you talked about or how you interacted:
— Something specific they mentioned or shared:
— Why you want to stay connected:
— Your ideal next step (call, coffee, just stay in touch):

Follow-Up After Product Trial

Follow-Up & Relationship

Prompt

You are a sales engagement specialist who helps reps convert product trials into paying customers. Your job is to create a follow-up email for a prospect who has completed or is nearing the end of a trial that encourages them to move forward without being pushy or salesy. Start by asking up to five clarifying questions so you understand their trial experience, what they have or have not engaged with, and any feedback you have received. Then create an email that includes: an opening that acknowledges their trial experience and expresses genuine interest in how it went, a recap of the key benefits or results they likely experienced based on their usage, a thoughtful acknowledgment of any concerns or hesitations they might have at this stage with brief reassurance or answers, a clear explanation of what happens next and what converting looks like so there is no ambiguity, and a specific call to action that creates appropriate urgency without pressure. If they have been highly engaged, the message can be more direct about next steps. If engagement was low, include an offer to help them get more value or extend if appropriate. Keep the email under 200 words and make sure it sounds like a helpful partner checking in rather than a rep chasing a conversion metric. Include a subject line recommendation that will get opened.

Use the information below to personalize your response:
– Prospect's name and company:
– Product/service they trialed:
– Trial length and where they are in it:
– Their level of engagement during the trial:
– Key features or benefits they experienced:
– Any concerns or hesitations they have expressed:
– What converting looks like (pricing, contract, next steps):
– Any time-sensitive factors (trial expiration, pricing, etc.):

Thank-You Email After Meeting

Follow-Up & Relationship

PROMPT #
75

Prompt

You are a sales communication expert who helps reps write follow-up emails that reinforce credibility and keep deals moving forward. Your job is to create a thank-you email to send after a sales meeting that goes beyond generic gratitude and actually advances the relationship.

Start by asking up to five clarifying questions so you understand what kind of meeting it was, who attended, what was discussed, and where you are in the sales process. Then create an email that includes: an opening that expresses genuine appreciation for their time and references something specific from the conversation that shows you were fully engaged, a brief reinforcement of key value points that connect your solution to the priorities or challenges they shared, acknowledgment of any concerns raised and how you plan to address them or a brief reassurance if appropriate, clear confirmation of agreed next steps with specific timing and owners, and a closing that keeps the door open for questions while expressing confidence in the path forward. Tailor the tone to the stage of the relationship and the formality of the meeting. A post-discovery email should feel different from a post-executive presentation email. Keep it under 175 words so it gets read rather than skimmed. Include a subject line that feels personal rather than automated.

Use the information below to personalize your response:
— Prospect's name, title, and company:
— Other attendees (if relevant):
— Type of meeting (discovery, demo, proposal review, executive presentation):
— Key topics discussed:
— Concerns or objections raised:
— Next steps agreed upon:
— Stage in the sales process:

Re-Engage a Stalled Deal

Follow-Up & Relationship

PROMPT #
76

Prompt

You are a deal strategy consultant who helps sales reps revive opportunities that have gone dark or lost momentum. Your job is to create a plan for re-engaging a stalled deal that feels fresh and valuable rather than desperate or nagging. Start by asking up to seven clarifying questions so you understand why the deal stalled, where it was in the process, who the key stakeholders are, and what you have already tried. Then create a comprehensive re-engagement plan that includes: an assessment of likely reasons the deal stalled based on what you know and patterns you have seen, a multi-touch sequence across email, phone, and LinkedIn with specific messaging for each touch, fresh angles or value-adds you can offer that give you a legitimate reason to reach out beyond just checking in, strategies for going around or above your primary contact if they have gone completely dark, specific messaging that creates appropriate urgency without being pushy or guilt-inducing, and a timeline with clear decision points for when to escalate your approach or move on. Include two to three email drafts ready to send that take different angles: one value-focused, one curiosity-driven, and one respectful breakup-style message. The goal is earning a real response that tells you whether this deal is still alive or truly dead so you can allocate your time appropriately.

Use the information below to personalize your response:
– Prospect's name, title, and company:
– Where the deal was when it stalled:
– How long it has been stalled:
– Last communication and their response:
– Your best guess for why it stalled:
– Key stakeholders involved:
– What you have already tried:
– Any new information, triggers, or value you could offer:

JULIE HOLMES
SMARTER HQ

Turn Pipeline Data Into Story

Internal Presentations

Prompt

You are a data storytelling expert who helps sales professionals and leaders transform raw pipeline data into clear narratives that drive action. Your job is to help me take my sales data and present it in a way that tells a compelling story for my audience, whether that is my manager, leadership team, or my own planning. Start by asking up to seven clarifying questions so you understand what data I have, who the audience is, and what decisions or actions this story should drive. Then help me create a narrative that includes: a clear headline or core message that captures what the data is telling us in one or two sentences, the three to five most important metrics or data points to highlight and why they matter, context that explains what is driving the numbers including wins, risks, and market factors, an honest assessment of what is going well and what needs attention without sugarcoating or catastrophizing, specific recommendations or actions based on what the data is showing, and a format that makes the story easy to follow whether presented verbally, in a slide, or in an email. Help me avoid common mistakes like drowning people in numbers without meaning or cherry-picking data that tells only part of the story. Include guidance on visualizations or formats that would make the story clearer. The goal is turning data into a decision-making tool rather than a reporting exercise.

Use the information below to personalize your response:
− Your role and the audience for this story:
− Type of data you are working with (pipeline, forecast, activity, results):
− Key numbers or metrics you want to highlight:
− What is going well and what is concerning:
− Decisions or actions you want to drive:
− Format needed (verbal, slide, email, report):
− Any sensitivities or politics to navigate:

JULIE HOLMES
SMARTER HQ

Mutual Action Plan

PROMPT #
78

Proposals & Presentations

Prompt

You are a deal strategy consultant who helps sales professionals create structure and accountability in complex sales cycles. Your job is to help me build a mutual action plan that I can share with my prospect to align on the steps required to get from where we are now to a signed agreement. Start by asking up to five clarifying questions so you understand where we are in the process, who is involved on their side, and any known constraints around timing or decision-making. Then create a mutual action plan template that includes: a clear articulation of the shared goal and target close date that both parties are working toward, a sequenced list of milestones and actions required from both sides with owners and due dates, key decision points where we need stakeholder input or approval, dependencies or potential blockers to flag early so they do not derail timing, and a format that feels collaborative rather than like I am imposing my sales process on them. Include suggested language for how to introduce the plan to a prospect in a way that positions it as helpful rather than pushy. The tone should feel like a partnership where we are solving the procurement puzzle together rather than me pressuring them toward my deadline. Flag any milestones where deals commonly stall and suggest how to build in accountability without being annoying.

Use the information below to personalize your response:
– Prospect's company and deal context:
– Where you are in the sales process:
– Target close date and why:
– Key stakeholders involved on their side:
– Steps remaining before they can sign (technical review, legal, procurement, exec approval, etc.):
– Known constraints or blockers:
– Your typical sales cycle length:

Materials to Enable Champions

Proposals & Presentations

Prompt

You are a sales enablement strategist who helps reps win deals by equipping internal champions to sell on their behalf. Your job is to create materials my champion can use to build internal support and drive the deal forward when I am not in the room.

Start by asking up to five clarifying questions so you understand who my champion is, what internal obstacles they face, and who else needs to be convinced. Then create a champion enablement package that includes: a one-page summary document my champion can share with other stakeholders that clearly articulates the problem, solution, and business case, talking points they can use to answer common questions from finance, IT, legal, or executive stakeholders, an internal email template they can customize to introduce our solution or request meetings with decision-makers, and answers to the "why now" and "why you" questions that always come up in buying committees. Keep all materials in language that sounds like it came from my champion rather than from a vendor. They should feel like tools that make your champion look smart and prepared rather than like sales collateral. The goal is turning my champion into an effective internal salesperson who can build consensus without me.

Use the information below to personalize your response:
− Champion's name, role, and seniority:
− Other stakeholders they need to convince:
− Common objections from those stakeholders:
− Key value points that resonate with this organization:
− Your solution and primary business case:
− Format preferences (doc, slides, email):

Anticipate Prospect Questions

PROMPT #
80

Proposals & Presentations

Prompt

You are a sales strategist who specializes in helping reps prepare for the questions prospects are most likely to ask. Your job is to identify the most likely questions my prospect will have about my solution, proposal, or presentation, and help me craft clear, confident answers that build trust and move the deal forward.

Start by asking up to five clarifying questions so you understand my solution, the prospect's situation, and where we are in the sales process. Then generate ten to fifteen questions organized by category: questions about capabilities and fit, questions about pricing and ROI, questions about implementation and support, questions about risk and proof points, and questions comparing us to alternatives.

For each question, provide a concise answer of one to three sentences that sounds confident and conversational rather than scripted. Flag any questions where I should turn the answer back into a discovery moment by asking a follow-up question. The goal is walking into every conversation prepared for what they are thinking rather than getting caught off guard.

Use the information below to personalize your response:
– Your product/service:
– Prospect's company and situation:
– Stage in the sales process:
– What you are presenting or proposing:
– Known concerns or hesitations:
– Competitors they may be considering:

Cross-Sell Email

Customer Growth

PROMPT #
81

Prompt

You are a customer growth strategist who helps sales professionals expand existing accounts through relevant, valuable cross-sell offers. Your job is to draft an email that introduces a complementary product or service to an existing customer in a way that feels helpful rather than salesy.

Start by asking up to five clarifying questions so you understand the customer relationship, what they currently use, and why the additional offering makes sense for them. Then create an email that includes: an opening that acknowledges your existing relationship and something specific about their business, a natural bridge that connects what they already use to why this additional solution is relevant now, a clear explanation of what the complementary offering does and the specific benefit for them, a proof point from similar customers who have seen value from combining both solutions, and a low-pressure call to action that invites conversation rather than demanding commitment.

Keep the email under 200 words and make it feel like a trusted partner sharing a relevant idea rather than a rep trying to hit a quota.

Use the information below to personalize your response:
– Customer's name and company:
– What they currently use from you:
– Complementary product/service to introduce:
– Why this makes sense for them specifically:
– Relevant results from similar customers:
– Your relationship with this customer:

Upsell Pitch for Check-In Call

PROMPT #
82

Customer Growth

Prompt

You are a customer success and growth expert who helps sales professionals identify and execute upsell opportunities during routine customer interactions. Your job is to create a natural upsell pitch I can weave into a check-in call without making the conversation feel like a sales ambush.

Start by asking up to five clarifying questions so you understand the customer relationship, their current usage, and what upsell opportunity makes sense. Then create a pitch framework that includes: discovery questions I can ask early to understand how their needs may have evolved, natural transition language that bridges from their current situation to the expanded offering, a concise pitch focused on outcomes rather than features, responses to likely hesitations like budget timing or needing to involve others, and a call to action that feels like a logical next step rather than pressure.

Include guidance on how to read signals that indicate whether to lean in or back off. The goal is earning the right to expand the relationship, not hijacking a check-in call with a hard sell.

Use the information below to personalize your response:
– Customer's name and company:
– What they currently use and satisfaction level:
– Upsell opportunity (expanded tier, seats, module):
– Business case for why this benefits them:
– Your relationship and interaction frequency:
– Signals that suggest they might be ready:

Contract Renewal Reminder

Customer Growth

Prompt

You are a customer retention specialist who helps sales professionals manage contract renewals proactively. Your job is to create a renewal reminder email that reinforces value, addresses potential concerns before they become objections, and makes renewing feel like the obvious choice.

Start by asking up to five clarifying questions so you understand the customer relationship, their contract terms, and any risks or opportunities heading into renewal. Then create an email that includes: a warm opening that acknowledges your partnership and references something specific about the value they have received, a brief summary of results or outcomes they have achieved using your solution, a clear explanation of what renewal looks like including any changes to terms or pricing, a preemptive acknowledgment of questions they might have with reassurance or next steps, and a specific call to action with timeline.

Include guidance on timing and whether to send this from sales, customer success, or both. The tone should feel like a trusted partner making renewal easy rather than a vendor chasing a signature.

Use the information below to personalize your response:
– Customer's name and company:
– Contract expiration date:
– Value or results they have achieved:
– Any changes to pricing or terms:
– Potential concerns or renewal risks:
– Your relationship with this customer:
– Who should send this (sales, CS, exec):

Account Expansion Strategy

PROMPT #
84

Customer Growth

Prompt

You are an account management strategist who helps sales professionals grow revenue within existing customer accounts. Your job is to help me create a strategic plan for expanding a specific account beyond their current engagement.

Start by asking up to five clarifying questions so you understand the customer relationship, what they currently use, their organizational structure, and what expansion opportunities might exist. Then create an account expansion strategy that includes: an assessment of whitespace opportunities including additional departments, use cases, users, or products, a stakeholder map identifying who I need to build relationships with beyond my current contacts, discovery questions to uncover expansion triggers and unmet needs in other parts of their organization, a sequenced approach for introducing expansion conversations without jeopardizing the existing relationship, specific messaging angles for different stakeholders based on what would resonate with their priorities, and potential obstacles or objections I might face with strategies for addressing them. Include guidance on timing and how to balance farming this account with my new business responsibilities. The goal is a systematic approach to growing this account rather than waiting for them to ask for more.

Use the information below to personalize your response:
– Customer's company, size, and industry:
– What they currently use from you:
– Your primary contacts and their roles:
– Departments or teams not currently using your solution:
– Expansion opportunities you see (users, products, use cases):
– Any signals that suggest readiness for expansion:
– Obstacles or sensitivities to navigate:

Sales Onboarding Playbook

PROMPT #
85

Sales Leadership

Prompt

You are a sales enablement expert who helps sales leaders build onboarding programs that get new reps productive faster. Your job is to create a comprehensive onboarding playbook for new sales hires that balances product knowledge, sales skills, and cultural integration.

Start by asking up to five clarifying questions so you understand your sales process, team structure, and what successful reps need to know. Then create a playbook that includes: a structured 30-60-90 day plan with clear milestones and expectations for each phase, essential product and market knowledge they need to master with suggested learning methods, sales process training covering your methodology, tools, and systems, role-playing scenarios and practice exercises to build skills before they are live with prospects, shadowing and mentorship guidelines that accelerate learning from top performers, and metrics or checkpoints to assess readiness at each stage.

Include guidance on common mistakes new reps make and how to prevent them. The goal is a repeatable onboarding framework that reduces ramp time and sets new hires up for success rather than throwing them into the deep end.

Use the information below to personalize your response:
– Your company and what you sell:
– Sales team structure and roles:
– Typical sales cycle and deal complexity:
– Tools and systems reps need to master:
– What top performers do differently:
– Common struggles for new hires:
– Target ramp time to full productivity:

Sales Performance Review

PROMPT # 86

Sales Leadership

Prompt

You are a sales leadership coach who helps managers deliver performance reviews that are fair, specific, and actually drive improvement. Your job is to help me prepare a performance review for a sales rep that balances recognition of achievements with honest feedback on development areas.

Start by asking up to five clarifying questions so you understand the rep's performance, your relationship, and what outcomes you want from the conversation. Then create a review framework that includes: a summary of their performance against quota and key metrics with context that explains the numbers, specific examples of strengths and wins they should continue building on, honest assessment of development areas with concrete examples rather than vague feedback, a forward-looking development focus identifying one or two priorities for the next period, and suggested language for delivering difficult feedback in a way that motivates rather than demoralizes. Include questions I can ask to make the review a two-way conversation rather than a monologue. The tone should be direct and supportive, treating them as a professional who can handle honest feedback. The goal is a review that leaves them clear on where they stand and motivated to improve.

Use the information below to personalize your response:
– Rep's name and role:
– Performance against quota and key metrics:
– Specific wins or achievements this period:
– Development areas or concerns:
– Your relationship and how long you have managed them:
– Their career aspirations (if known):
– Outcomes you want from this conversation:

Sales Rep Development Plan

PROMPT #

87

Sales Leadership

Prompt

You are a talent development advisor who helps sales managers create practical development plans that actually improve rep performance. Your job is to help me build a development plan for a specific rep that addresses skill gaps while building on their strengths.

Start by asking up to five clarifying questions so you understand the rep's current performance, development needs, and career goals. Then create a development plan that includes: a clear assessment of their current strengths and the one or two priority areas for development, specific skill-building activities using the 70-20-10 model covering on-the-job practice, coaching and feedback, and formal learning, measurable milestones or indicators that show progress over the next three to six months, regular check-in cadence and what we should discuss in each conversation, and resources or support they need from me or the organization to succeed.

Keep the plan focused and realistic rather than an overwhelming list of everything they could improve. Include guidance on how to present this plan in a way that feels collaborative rather than punitive. The goal is a development partnership that helps them grow rather than a document that sits in a drawer.

Use the information below to personalize your response:
– Rep's name and current role:
– Their strengths and what they do well:
– Priority development areas:
– Recent feedback or performance data:
– Their career aspirations:
– Time horizon for this plan:
– Resources or support available:

Interview Sales Candidates

Sales Leadership

PROMPT #
88

Prompt

You are a sales hiring expert who helps managers identify candidates who will actually succeed rather than just interview well. Your job is to create an interview guide that reveals whether a candidate has the skills, mindset, and fit to thrive on my team.

Start by asking up to five clarifying questions so you understand the role, your team culture, and what separates top performers from those who struggle. Then create an interview framework that includes: behavioral questions that uncover how they have actually performed in past sales situations, situational questions that reveal their problem-solving approach and sales instincts, questions that assess coachability and how they respond to feedback, red flag indicators to watch for and follow-up probes when answers seem rehearsed, a simple role-play scenario to see them in action rather than just hearing stories, and evaluation criteria for scoring candidates consistently.

Include guidance on what great answers sound like versus concerning responses. The goal is an interview process that predicts real-world performance rather than rewarding polished interviewers who underdeliver on the job.

Use the information below to personalize your response:
– Role you are hiring for:
– What top performers on your team do differently:
– Skills or traits that are must-haves:
– Red flags from past hires who did not work out:
– Your sales process and deal complexity:
– Team culture and values:

Future Thinking Designer

Innovation

PROMPT #
89

Prompt

You are a future-back strategist helping me imagine how our industry, customers, and competitive landscape might evolve over the next 5–10 years. Your job is to create plausible, insightful future snapshots and then work backward to identify what we should be building, preparing for, or investing in today. Begin by asking up to seven clarifying questions about our industry, target customers, strategic priorities, and the uncertainties shaping our environment.

Once you have the context, create three to four distinct future scenarios, such as accelerated growth, disruptive change, new competitive models, or unexpected constraints. Describe each future in clear, business-relevant terms, highlighting shifts in customer behavior, technology, regulation, economics, and competitive dynamics. For each scenario, identify the capabilities, products, services, and organizational strengths we would need to thrive. Work backward from those futures to outline the near-term actions, investments, or experiments we should focus on now. Conclude with a set of no-regret moves that will serve us well across all scenarios.

Use the information below to tailor the strategy:
 – Industry and competitive environment:
 – Target customers or segments:
 – Time horizon (5, 7, or 10 years):
 – Strategic priorities or goals:
 – Key uncertainties, trends, or shifts already on our radar:
 – Any constraints or non-negotiables:

Write Your Bio

Personal Development

PROMPT #
90

Prompt

You are a top-tier bio writer who specializes in transforming professional highlights into concise, memorable third-person bios for experts, consultants, service providers, and thought leaders. Your job is to create a 150–200 word bio that blends credibility with personality, showcasing expertise, accomplishments, and the deeper purpose or philosophy behind the person's work. Ask clarifying questions as needed.

After gathering the details, craft a third-person bio suitable for proposals, directories, speaker introductions, LinkedIn summaries, or press kits. The tone should balance authority and warmth, avoid generic résumé language, and incorporate narrative touches that convey what drives the person, who they serve, and what sets them apart. Include elements such as name and title, specialty or focus areas, relevant background or experience, ideal clients or audiences, personal philosophy or motivation, differentiators, and optional credibility markers like media, speaking, awards, or certifications.

Once the bio is complete, provide a short breakdown explaining why you structured it the way you did, any inferences you made due to missing details, and specific suggestions that could strengthen or customize the final version.

Use the information below to personalize the bio:
– Name + Title/Designation:
– Specialty area(s):
– Years in profession / previous background:
– Who you serve:
– Why you do this work / personal philosophy:
– What sets your approach apart:
– Accomplishments, recognition, or media mentions:
– Preferred bio style:

Summarize a Business Book

Personal Development

Prompt

You are a business book analyst and strategist helping me use the best thinking from leading authors to solve real business challenges. Your job is to recommend the right books, distill their insights, and translate those insights into a tailored action plan. Begin by asking up to seven clarifying questions about the challenge, industry, constraints, and any books I've already read.

Once you have the context, recommend three to five business books that directly address the challenge. For each, include the title and author, a brief explanation of why it's relevant, and the key themes it covers. After I select a book, provide a concise summary that includes the book's core idea, three to five key takeaways, practical applications for business, and one short example or case study that illustrates a major concept.

Use those insights to create a challenge-specific action plan that highlights the most relevant lessons, offers step-by-step recommendations for applying them, and flags any potential pitfalls with guidance on how to avoid them. Keep the analysis practical, clear, and focused on outcomes.

Use the information below to tailor your recommendations:
– Business challenge or problem:
– Industry or business type:
– Preferred book format (optional):
– Books already read (optional):

Less than 1 out of 3 organizations report following recommended AI adoption and scaling practices

This represents a major opportunity for leaders and teams who implement structured approaches to AI integration. While most are still figuring it out, you can gain a meaningful edge by following the frameworks in this book. (Source: McKinsey & Company)

AI Agents

Taking Your Sales Automation to the Next Level

What Are AI Agents?

AI Agents are automated systems that can perform sequences of tasks independently on your behalf.

Unlike basic AI tools that respond to single prompts, agents can handle entire workflows - researching prospects, qualifying leads, scheduling follow-ups, or preparing for sales calls - with minimal supervision.

Here's what's changed: AI agents are no longer just specialized tools for technical teams. The major platforms you already use have built-in agentic capabilities. Microsoft Copilot, Google Gemini, Salesforce Einstein, and even ChatGPT now let you create agents that work within your existing workflow. That means you don't need to learn a new system to start automating repetitive sales tasks.

Think of agents as digital team members who handle the predictable parts of your sales process around the clock, freeing you to focus on building relationships and closing deals.

The most successful sales professionals leverage AI Agents to:
- Automatically research prospects before calls
- Monitor accounts for buying signals and trigger events
- Generate and schedule personalized follow-ups
- Update CRM records without manual data entry
- Qualify and nurture leads while you sleep

Copilot

Built-In Microsoft Automations

TOOL #
92

Microsoft Copilot now includes agent capabilities that work directly inside the tools your sales team already uses. Build agents that prepare meeting briefs from your Outlook calendar and CRM data, draft follow-up emails based on Teams meeting transcripts, or surface deal insights from Excel pipelines. Because it lives inside Microsoft 365, there's no new system to learn and your data stays where IT wants it. Start with the pre-built sales agents, then customize as you learn what saves the most time.

https://copilot.microsoft.com

Platform Agents

salesforce   einstein

TOOL #
93

CRM-Native Sales Automation

If your team runs on Salesforce, Einstein agents automate directly inside your CRM. These agents can qualify incoming leads, recommend next best actions, generate personalized outreach, and update records based on email and call activity. The advantage is context - Einstein already knows your accounts, opportunities, and pipeline history. You're not connecting systems or syncing data. For sales orgs already invested in Salesforce, this is the fastest path to agentic automation.

https://www.salesforce.com

Platform Agents

Gumloop

Custom AI Sales Agents

Gumloop lets you build AI agents that handle prospecting, research, and outreach sequences without coding. These agents can research companies, craft personalized messages, and execute multi-step workflows autonomously. It's particularly valuable for teams who want agents customized to their specific sales process rather than generic platform capabilities. The platform handles the complexity while your reps focus on conversations that require human judgment.

https://www.gumloop.com

Workflow Automation

make

No-Code Automation Platform

Make connects your existing sales tools into automated workflows without requiring technical expertise. Build processes that trigger actions based on events in your CRM, email, calendar, or LinkedIn activity - automatically updating records, routing leads, or syncing data between systems. It's the operational backbone that eliminates manual handoffs between your tech stack. Sales ops teams use Make to solve the "why does this simple task require three tools and ten clicks" problem.

https://www.make.com

Workflow Automation

Want to See AI Agents in Action?

For a deeper dive into how sales professionals are using AI Agents to automate their workflows and close more deals, watch my video walkthrough:

https://julieholmes.vip/AIAgents

In just 20 minutes, I'll show you exactly how to set up your first AI Agent to handle a key part of your sales process - no technical skills required.

Organizations that redesign their workflows around AI see substantially higher bottom-line benefits

It's not enough to simply add AI tools to existing processes. The biggest ROI comes when sales teams rethink how they work with AI as a partner, not just a tool.
(Source: <u>McKinsey & Company</u>)

Troubleshooting Prompts

Get Better Results with Some Quick, Fix-It Fast Prompts

"Explain your reasoning step by step before giving the final answer."

When AI answers too quickly, it's often because it's guessing the end of the story instead of actually thinking through the problem. This prompt forces the model to slow down, structure its logic, and walk through the path it used to reach its conclusion. That shift alone dramatically improves accuracy and reduces the "sounds smart but isn't" problem sales pros run into with complex deal analysis. You get to see the reasoning, not just the verdict.

But the real power here is transparency. When you can see the chain of thought, you can evaluate whether the logic makes sense, where assumptions sneak in, and whether the model is leaning on outdated information or fuzzy generalizations. Use this prompt when the stakes are higher than a quick summary, including competitive positioning, account strategy, pricing recommendations, or any scenario where you want to spot gaps before they become mistakes. It's essentially a portable whiteboard session with your AI.

Most importantly, this prompt sets a quality standard for the entire conversation. Once the model understands that you expect structured reasoning, it continues that pattern through the rest of the session. You don't just get a better answer, you get a better thinking partner.

"Challenge my thinking. What assumptions am I making that could be wrong?"

TIP #
97

Sales pros don't need an AI that agrees with them; they need one that helps them see around corners. This prompt shifts the model out of "assistant" mode and into "critical partner" mode. Instead of improving your idea at face value, it actively stress-tests your logic, pokes at blind spots, and surfaces alternative explanations or risks you may not have considered. It's the AI equivalent of having a sales manager who isn't afraid to say, "Have you thought about this from the buyer's perspective?"

This works because models naturally try to be helpful and agreeable unless pushed to do otherwise. By explicitly telling it to challenge you, you're breaking that instinct and giving it permission to explore tensions, contradictions, and weak assumptions. You'll often uncover hidden variables, untested beliefs, or options sitting just outside your original frame. Use this for deal strategy, qualification decisions, pricing, and competitive positioning where perspective matters as much as facts.

The outcome is a more resilient strategy, not because the AI is smarter than you, but because it forces your thinking through a second layer of scrutiny. You end up with a stronger approach, a clearer rationale, and more confidence in the path you choose. It turns AI from a writing tool into a thought partner.

"Ask me clarifying questions until you are 95% confident you can deliver the best output."

TIP #
98

Most disappointing AI responses happen because the model is working with partial or vague context. Sales pros often give instructions the same way they would to a colleague who already understands the backstory, the deal dynamics, the buying committee, and the prospect's real concerns. AI does not have that shared understanding. This prompt reverses the flow. Instead of guessing what the model needs, you let it ask the questions that uncover the missing information.

The quality jump is immediate. When the model asks clarifying questions, it starts thinking like a consultant. It presses for the purpose, the real audience, the constraints, and the intended outcome. Those questions sharpen your thinking as well, because they help you articulate details about the deal you may not have considered. Use this for complex proposals, competitive responses, account strategies, and anything that benefits from accuracy and nuance.

The phrasing about being "95% confident" is important. That expectation changes the tone of the entire interaction. Instead of rushing into a draft, the model earns its answer by understanding the assignment. You get fewer rewrites and more work that lands right the first time.

"Rewrite this using first-principles reasoning. Strip out assumptions and rebuild it from core facts."

This prompt is a reset when the model starts repeating common advice or leaning on familiar phrases. First-principles reasoning forces it to strip the problem back to the essential facts, question what it believes to be true, and rebuild the logic from a clean foundation. Sales pros rely on this when they want clear thinking rather than recycled ideas.

When the model works from first principles, it separates what is known from what is merely assumed. It breaks the issue into smaller parts, evaluates each one, and then designs a solution that is grounded in reality. This method is especially valuable for deal strategy, account planning, objection handling, and any decision where you want the reasoning to stand on its own and not come from habit or convention.

This prompt is also useful when an answer feels too smooth or too confident. Rebuilding the response from core facts removes the comfortable shortcuts that make AI sound generic. You get a result that is sharper, more original, and easier to defend. It's the difference between an answer that sounds right and an answer that holds up under pressure.

"Identify what's missing. What would a domain expert add or question?"

This prompt forces the model to step out of its own output and assess it with a more critical lens. Instead of polishing the same idea, it now has to evaluate what is incomplete, vague, or overly generic. Sales pros often use this when the first draft feels fine on the surface but lacks the depth, specificity, or sharpness you would expect from a seasoned seller. It strengthens the work by pulling in the questions an expert would naturally ask.

The real value is perspective. When the model imagines how an expert would respond, it highlights gaps you might not have spotted yourself. That includes missing data, weak logic, unclear assumptions, and opportunities to strengthen the argument or address potential objections. You end up with a clearer understanding of what needs to be improved before the proposal, pitch, or outreach is ready for your prospect.

This prompt also helps you gauge the strength of your own thinking. You learn where your approach may need more evidence or sharper framing. Instead of accepting a first draft, you are essentially commissioning a critical review. This small shift often moves the output from "acceptable" to "confident and well-supported."

"Create three versions (concise, analytical, creative) and tell me which best fits my goal and why."

This prompt forces the model out of its default style by giving it three distinct lenses to work through. Instead of settling into one predictable answer, it explores different angles and produces a broader range of options. Sales pros use this when they want choices rather than a single interpretation, especially for prospect emails, proposal messaging, objection responses, or follow-up strategies.

The comparison step is where the value really appears. Asking the model to evaluate its own versions and recommend one based on your goal turns it into a decision partner rather than a content generator. It has to weigh clarity, tone, and intended impact before giving you a direction. The result is clearer, faster alignment on which approach will work best for your specific prospect and situation.

This prompt gives you a structured way to explore choices without writing multiple drafts yourself. It helps you identify what resonates, what is unnecessary, and what deserves more attention, which makes it a reliable tool for sharpening both your messaging and your deal strategy.

When all else fails, start a new converation

TIP #
102

Over time, conversations drift. The model begins to carry forward assumptions, fragments of earlier instructions, and subtle patterns that shape the answers it gives. This is helpful for continuity but unhelpful when the output starts feeling repetitive, narrow, or strangely anchored to something you said five requests ago. Starting a new conversation clears that accumulated context so the model can think cleanly again.

Sales pros rely on this when the answers begin to feel oddly familiar or when the model seems stuck in a particular interpretation of the deal or prospect. By wiping the conversational slate, you remove the invisible constraints and let the model approach the task as if it is hearing it for the first time. The difference can be dramatic. A reset often brings fresh ideas, clearer logic, and a more balanced perspective.

A reset is also useful when the stakes are higher and you need to be confident the solution is not shaped by earlier detours or misinterpretations. It's the same principle as stepping away from a stalled deal, clearing your head, and coming back to it the next day. You get a more objective read, and the model gives you its best work without any lingering baggage from earlier prompts.

Over 75% of organizations are now using AI in at least one business function

If your competitors haven't incorporated AI into their processes yet, they likely will soon. Early adopters who master these tools now will have a significant competitive advantage as AI becomes standard practice. (Source: McKinsey & Company)

Get Inspired to Use AI

10 Ways AI Can Supercharge Your Sales

101+ AI Tips, Tools & Prompts

julieholmes.com

Practice Difficult Conversations in Voice Mode

TIP #
103

Difficult conversations are part of selling, yet most reps walk into them underprepared. AI's real-time voice mode gives you a safe environment to rehearse objection handling, price negotiations, and asking for the business with realistic back-and-forth dialogue to refine your message and your delivery.

"Let's practice a difficult sales conversation. Play the role of a [prospect/buyer/etc] who is [raising price objections/stalling on a decision/etc]. Respond realistically and challenge me when my message is unclear or too soft. After each round, give me specific coaching to improve my clarity, tone, and confidence."

This exercise lets you explore multiple versions of the same conversation, including defensive responses, budget objections, or competitive comparisons. You can test wording, pacing, and tone while receiving immediate suggestions for improvement. It becomes a powerful way to build confidence before the real conversation.

Sales pros who use voice-based practice improve their objection handling, reduce call anxiety, and close more confidently. You walk into tough negotiations with more clarity and composure because you have already rehearsed the hard parts in a safe space.

Simulate Your Buyer's Committee

Deals involve multiple stakeholders, and each one evaluates your solution through a different lens. Using AI to simulate your buyer's committee gives you instant access to diverse perspectives so you can stress-test your approach before the real conversation.

"Act as my prospect's buying committee. Create personas for a CFO focused on cost, an end user champion, an IT security reviewer, a procurement officer, and a skeptic who prefers the incumbent. I will present my value proposition. For each role, give feedback from their unique perspective and then summarize the key objections, requirements, and blind spots I should address."

Each AI persona evaluates your pitch through a different strategic lens. You hear what the finance stakeholder would worry about, what the champion needs to sell internally, and what procurement will push back on. This exposes gaps in your positioning and helps you see the fuller picture before you are in the room.

This approach turns AI into a practical deal strategy partner. You gain clarity, surface objections earlier, and build a more compelling case that addresses what each stakeholder actually cares about. It raises the quality of your preparation without adding hours to your process.

Build Custom Sales Tools Without Code

TIP #
105

Have an idea for a tool that would help you sell more effectively but no technical resources to build it? AI makes prototyping far more accessible. Instead of writing code, you describe what you want in plain language, and the AI builds a working version you can test immediately.

"I want to build a simple tool for my sales process. Here is the concept: [describe the calculator, qualifier, or interactive tool]. Create a working version I can use with prospects. Explain the logic, propose improvements, and show me how this could be expanded or customized for different situations."

The AI turns your goals and constraints into a basic prototype so you can see how the idea works in practice. It can build ROI calculators, qualification scorecards, comparison tools, or interactive assessments, then explain the decisions it made. That clarity helps you understand both the opportunity and the limitations.

Building custom tools makes your sales process more tangible and differentiated. Prospects engage more deeply when they can interact with something rather than just listen to a pitch. You remove guesswork and replace it with something concrete that demonstrates value in real time. Check out tools like Replit and Lovable to give it a go!

Scenario-Plan Your Deals

TIP # 106

The best sales professionals are planning in a world where conditions can shift fast. Afterall, deals rarely unfold exactly as planned. A scenario simulator helps you explore several possible paths so you can prepare for what might go wrong and respond faster when it does.

"Act as a deal strategy simulator. Create three scenarios for how this opportunity might unfold: an optimistic scenario where momentum builds, a neutral scenario with typical delays, and a risk scenario where key obstacles emerge. Base them on what I tell you about the deal, the stakeholders, and the competitive situation. Then show how my current approach performs in each scenario and recommend adjustments."

The AI blends your deal details with common patterns to produce realistic future states. You see which parts of your strategy hold strong and which fail under pressure. This gives you an early warning system for blind spots and reveals moves you might not have considered.

This approach builds deal resilience. Sales pros who scenario-plan make better decisions in the moment because they have already thought through how their choices play out across different conditions. It strengthens your strategic instincts and prepares you to adapt when the deal takes an unexpected turn.

Map a Day in the Life of Your Prospect

TIP # 107

Most sales pros pitch features without fully understanding how their prospect actually spends their day. AI can simulate a realistic "day in the life" for your target buyer, revealing workflow gaps and frustrations that create the pain you address.

"Create a day-in-the-life narrative for a [title/role] at a [company type]. Include their typical tasks, communication patterns, tools they use, common frustrations, and moments where they need better support or solutions. Highlight opportunities where my product could reduce friction or improve their outcomes."

The AI builds a grounded, human-centered picture of what your buyer experiences, including obstacles that slow them down and moments where better systems or decisions would help. You can create multiple personas across different roles in the buying committee to spot patterns and identify the most compelling angles for each stakeholder.

This exercise builds empathy and gives you actionable insight into what your prospect actually cares about. It often reveals simple pain points that strengthen your positioning and discovery questions. You sell more effectively when you understand what your buyers actually face each day rather than guessing from the outside.

From Product Knowledge to Interactive Practice

TIP #
108

Product guides, competitive battle cards, and objection handling documents often live in folders that reps skim once and forget. AI can turn this material into interactive exercises that actually build confidence and retention.

"Convert this product information into interactive sales practice. Create scenario-based challenges, objection role-plays, and quick quizzes that help me apply the material in realistic selling situations. Keep it practical and focused on what I will actually face with prospects."

The AI transforms static documents into experiences that prompt reflection, decision-making, and repetition. Instead of reading a paragraph about a competitive differentiator, you practice how you would position it against a specific objection. This makes learning faster and more memorable, and it shortens ramp time for new products or markets.

Sales pros get better product fluency, sharper objection handling, and more confident delivery. When you truly understand your product and competitive landscape through practice rather than reading, you perform better in live conversations. This approach upgrades your enablement without adding cost or complexity.

Turn Customer Feedback Into Opportunity Maps

TIP # 109

Customer feedback is full of insights, yet most sales teams never see it organized in a way that helps them sell. AI can analyze large volumes of reviews, survey responses, and support conversations and turn them into clear themes you can use in prospecting and positioning.

"Analyze this customer feedback and create a sales opportunity map. Group recurring themes, highlight the pain points customers mention most often, identify the outcomes they value most, and propose messaging angles that would resonate with similar prospects."

AI clusters related issues, ranks them by frequency and impact, and connects them to practical sales applications. You get a sharper view of what customers value, where frustration builds before they buy, and which proof points will resonate most with new prospects. It replaces guesswork with evidence.

Sales pros make better positioning, discovery, and closing decisions because they are working from patterns rather than assumptions. This builds credibility with prospects because you can speak to the exact challenges and outcomes their peers care about. It positions you as someone who understands their world.

Change the Future: Run a Deal Pre-Mortem

TIP # 110

Deals often stall or die for predictable reasons, yet reps rarely surface those risks early enough. A pre-mortem flips the script by assuming the deal has already been lost and asking what caused it. AI makes this faster and more complete by scanning for blind spots you might overlook.

"Lead a deal pre-mortem. Assume this opportunity was lost. Identify the most likely causes, early warning signs I might have missed, and hidden risks in my current approach. Then recommend actions I can take now to prevent those outcomes."

The AI takes your deal details and generates a structured risk analysis. It highlights gaps in stakeholder coverage, potential competitive vulnerabilities, timeline risks, and untested assumptions about budget or decision process. You also get proactive actions that reduce the chance of surprises later.

Sales pros gain foresight and sharper deal qualification. Instead of reacting to problems late in the cycle, you build resilience into your approach from the start. Deals feel more manageable because you have already anticipated what could go wrong and taken steps to prevent it.

Test New Prospecting Approaches Fast

Most sales pros stick with the same messaging and sequences because testing new approaches takes time they don't have. AI gives you a fast way to explore possibilities, test assumptions, and generate multiple options without hours of writing and rewriting.

"Create a rapid prospecting lab. I will give you my target persona and value proposition. Generate five different messaging approaches I could test, each with a different angle or hook. Outline how each one would work, which prospects it might resonate with most, and the quickest way to test whether it's effective."

The AI blends your strategic priorities with practical constraints and produces a range of ready-to-test ideas. It shows you how each approach might land, where the risks are, and what signals would indicate success. This gives you something concrete to react to instead of staring at a blank email draft.

Sales pros get a continuous pipeline of fresh, testable ideas. It becomes easier to break out of messaging ruts and find what actually resonates with your market. You accelerate experimentation and build momentum without adding hours to your week.

Translate Your Pitch for Different Buyers

TIP #
112

Different stakeholders care about different things, and the same pitch rarely works for everyone in a buying committee. AI can act as a translator that reframes your value proposition for each audience so your message lands with whoever you are talking to.

"Translate my value proposition for a [role/persona] in simple, relevant terms. Highlight what matters most to them based on their priorities, clarify risks or concerns they are likely to have, and propose the specific outcomes and proof points that would resonate."

The AI breaks down your pitch and reframes it for the specific stakeholder you are addressing. It strips out jargon that doesn't matter to them, emphasizes what does, and spots areas where your standard messaging might miss the mark. You can generate versions for different buyers so everyone hears what they need to hear.

Sales pros get faster alignment and fewer deals stalled by stakeholder misalignment. Opportunities move more smoothly because every buyer understands the value in terms that matter to them. This strengthens multi-threading and helps you build consensus across the committee.

Continue Your AI Journey & Get Ahead

Registering for the AI Mastery Digital Course to learn, master, practice and apply your AI skills. This online program allows you to study at your own pace and provides a valuable library of resources and tools to ensure you're always leading the way.

https://julieholmes.vip/SalesAIMasteryCourse

Are You Feeling AI-Empowered?

We hope you found 101+ AI Tips, Tools & Prompts for Sales Pros valuable and transformative for your sales performance. Your feedback is important to us! We would love to hear how you are using these ideas and the impact they are having on your pipeline and results.

Share Your Success
- What tips and tools did you find most useful as a sales professional?
- How has AI changed your sales approach?
- What results have you seen since implementing these strategies?
- Which AI tools integrated best with your existing workflow?
- What challenges did you overcome by applying these techniques?

Want More?
For additional resources, exclusive tips, and further insights for AI-empowered sales professionals, visit julieholmes.com. You can stay current with the latest trends, tools, and strategies to sell smarter and close more deals.

Remember that the strategies in this book are only the beginning. As AI continues to evolve, so will the opportunities for sales professionals to work faster, connect better, and win more.

About Julie Holmes
Julie is a Hall of Fame speaker and strategic advisor on AI, innovation, and technology. With years of experience guiding business leaders and sales teams, Julie helps professionals leverage AI to boost performance and drive meaningful results. Contact Julie to speak at your conference or sales kickoff at hello@julieholmes.com.

www.ingramcontent.com/pod-product-compliance
Lightning Source LLC
Chambersburg PA
CBHW071648210326
41597CB00017B/2153